PHOTODIODES - COMMUNICATIONS, BIO-SENSINGS, MEASUREMENTS AND HIGH-ENERGY PHYSICS

PHOTODIODES - COMMUNICATIONS, BIO-SENSINGS, MEASUREMENTS AND HIGH-ENERGY PHYSICS

Editor

Eden Slaver

scitus academics

Photodiodes - Communications, Bio-Sensings, Measurements and High-Energy Physics
Edited by **Eden Slaver**

ISBN: 978-1-68117-599-7
Library of Congress Control Number: 2016946454

www.scitusacademics.com

Notice

Printed in United States of America on Acid Free Paper

PREFACE

A photodiode is a light detector type that consists of the light conversion process into voltage or may be present, depending on the operating mode of the unit. It is made up of optical filters and integrated lenses with large or small surfaces. Photodiodes have a reduced response time by increasing their surface area. Conventional photovoltaic cells used to generate electric solar energy are an ordinary, large-scale photodiode. A photodiode is a semi-conductor with a p-n interface and an intrinsic layer between p and n layers. It generates photocurrent by generating pairs of electron hole due to the absorption of light in the intrinsic or possibly depleted region. The photocurrent produced is proportional to the light intensity absorbed. The photodiode mechanism is like this in a (miniaturized) solar cell. The answer time is rapid in the order of nanoseconds. Although their linearity can be used in simple light meters, they are considerably less vulnerable as a photo transistor.

This book describes various types of photodiodes for high-speed data communication, UV-light detection, and high-speed measurement, biomedical sensing and excessive energy physics applications. Photodiodes are used in consumer electronics equipment such as compact disk players, smoke detectors and infrared remote control receivers used to control equipment from televisions to air conditioners. Photoconductors or photodiodes can be used for many applications. Either type of photosensor can be used to measure light, such as in camera light meters, or perhaps to respond to light levels, such as at night when the street light is switched on. Photosensor of all types can be used to respond to incident light or possibly to a light source which is part of the same system or circuit.

TABLE OF CONTENTS

1

GEIGER-MODE AVALANCHE PHOTODIODE ARRAYS INTEGRATED TO ALL-DIGITAL CMOS CIRCUITS

Brian Aull

Massachusetts Institute of Technology Lincoln Laboratory, 244 Wood St, Lexington, MA 02420, USA; Tel.: +1-781-981-4676

ABSTRACT

This article reviews MIT Lincoln Laboratory's work over the past 20 years to develop photon-sensitive image sensors based on arrays of silicon Geiger-mode avalanche photodiodes. Integration of these detectors to all-digital CMOS readout circuits enable exquisitely sensitive solid-state imagers for lidar, wavefront sensing, and passive imaging.

KEYWORDS

imagers; imaging; avalanche photodiodes; Geiger-mode avalanche photodiodes; lidar; ladar; wavefront sensing; photon counting

1. INTRODUCTION

The MIT Lincoln Laboratory (Lexington, MA, USA) has been a world leader in the development of specialized high-performance charge-coupled-device imagers (CCDs) [1]. CCDs have outstanding performance and support advanced functions such as charge-domain image stabilization, electronic shuttering, and blooming control. The ultimate sensitivity limitation of a CCD is set by the readout noise of the output amplifier that senses the charge packets and converts them to analog voltage levels. The faster the readout rate, the more severe the readout noise penalty. This sensitivity limitation becomes important in photon-starved applications, such as night vision or high-temporal-resolution imaging.

Interest in such scenarios lead to Lincoln's development of photon-counting image sensors, primarily based on arrays of custom-fabricated Geiger-mode avalanche photodiodes (GMAPDs) integrated with all-digital CMOS readout circuits. The term "photon counting" is used broadly here to mean that each photon arrival is digitally recorded by the pixel circuit. The pixel could be designed to either time stamp photons or count them. While the primary focus of this article is silicon GMAPDs [2], Lincoln Laboratory has developed GMAPD arrays based on compound semiconductors sensitive at wavelengths ranging from 1.06 μm [3] to the mid-wave infrared [4].

The principal advantage of a photon counting image sensor is that it eliminates readout noise. Because digitization occurs within the pixel, there is no analog circuitry in the readout path, and therefore no analog circuit noise. This means that there is no readout noise penalty for operating at high frame rates, using short integration times, or dividing the incoming light into multiple spectral or spatial channels. In-pixel digitization therefore enables high-frame-rate imaging [5,6], multispectral imaging [7], and spatial oversampling [8].

A second advantage of a photon-counting image sensor is that it facilitates in-pixel time-to-digital conversion. The ability to digitally time stamp individual photon arrivals is an enabler for exquisitely sensitive lidar imaging systems. The GMAPD arrays can also be used as laser communication receivers in systems with challenging link budgets. Use of pulse-position-modulated formats allows for multiple bits of information to be encoded in a single transmitted photon [8].

The integration of GMAPDs with digital CMOS can take advantage of Moore's Law scaling and 3D integration to incorporate on-focal-plane processing functions such as imaging stabilization, spatial filtering, change detection, and tracking of objects of interest without mechanical scanning.

Potential disadvantages of GMAPDs include afterpulsing and crosstalk. Afterpulsing denotes false detection events triggered by carriers generated in previous events. To avoid afterpulsing the avalanche photodiode (APD) must be debiased for a sufficient time after an event to give such residual carriers time to recombine or be collected. Since the APD is unresponsive during this time, this leads to blocking loss and causes the count rate to saturate at high optical fluxes. Crosstalk is mediated by hot-carrier light emission by the avalanche process and is discussed at the end of Section 4.

2. GEIGER-MODE APDS AND PHOTON-TO-DIGITAL CONVERSION

There are many types of detectors that have single-photon sensitivity, including photomultiplier tubes, image intensifiers, superconducting nanowires, and CCDs with avalanche gain registers [9]. GMAPDs, also known as single-photon avalanche detectors or SPADs [10], have a combination of advantages over these other photon-counting technologies. It is an all-solid-state device technology, scalable to many pixels, capable of room-temperature or thermoelectrically-cooled operation, and as noted already, amenable to on-focal-plane digitization and processing.

An avalanche photodiode is a *p-n* junction diode whose doping profile supports high electric fields near the junction at operational bias. A photoelectron or photohole created by the absorption of a photon is accelerated to sufficient energy to initiate a chain of impact ionization events, creating offspring electron-hole pairs and leading to internal gain. Once this avalanche process is underway, there is a competition between carrier generation and carrier extraction. Below the avalanche breakdown voltage, carriers are being extracted faster than they are being generated. The current flow decays, leading to self-termination of the avalanche process. This gives finite gain and therefore produces a photocurrent that is proportional to the intensity of the incident light. This is the traditional mode of operation used in optical communication receivers, known as linear mode. The gain is an increasing function of bias, and there is a random variation of gain that degrades signal-to-noise ratio.

The APD can also be operated above the avalanche breakdown voltage. When an avalanche starts, carrier generation predominates over extraction leading to exponential growth of the current. This growth is arrested by series resistance (in most cases mediated by space charge buildup). This mode of operation is known as Geiger mode, in which a single photon can initiate an avalanche that is self-sustaining. Of course, the APD must then be reset by reducing the bias to below breakdown long enough to terminate the avalanche, a process known as quenching.

The physics of Geiger-mode operation lends itself to a startlingly simple way of interfacing the GMAPD to a digital CMOS circuit: a direct connection to the input of a logic element, illustrated in Figure 1. The *p*-side of the APD is tied to a negative bias voltage slightly less in magnitude than the breakdown. The *n*-side is connected to a logic circuit that performs simple voltage sensing. Initially, the n-side of the APD is set to a logic-high voltage by briefly turning on a pull-up transistor. Once armed in this fashion, the APD has a reverse bias several volts above breakdown. When an avalanche is initiated, the APD turns on and then discharges its own parasitic capacitance. Once at or below breakdown, the avalanche terminates, leaving the *n*-side at a voltage close to a logic 0. The APD is effectively a CMOS-compatible digital element. Also, like a CMOS logic element, the APD draws current only when it is switching. This "photon-to-digital conversion" scheme has been the basis of all of Lincoln's GMAPD imaging devices, with minor modifications. One can, for example, add a pull-down transistor to speed up the APD discharge, size the transistors in the sensing gate to adjust the triggering threshold, or cascode transistors to augment the voltage swing. Voltage sensing gives more timing latency and timing jitter than current sensing circuits [11]. This drawback has been more than offset by the advantages of the photon-to-digital conversion scheme: simplicity, robustness, and freedom from static power dissipation.

3. FABRICATION AND INTEGRATION WITH CMOS

Lincoln Laboratory's silicon APD arrays are fabricated in house in its Microelectronics Laboratory, currently on 200-mm silicon substrates. Typically, the substrate is heavily p doped, with a lightly p-doped epitaxial layer on which the APD structure is fabricated. The doping profile of the APD is defined by a series of patterned ion implantation steps. After back-end processing to make contact pads, the APD array is integrated with a CMOS readout circuit, using either bump bonding or a 3D-integration technique. During this process the opaque silicon substrate must be removed to enable back-side illumination. The back side is passivated with a p-doped contact layer. This layer is typically implanted and then activated using laser annealing, although molecular beam epitaxy has also been used to grow the contact layer. In many cases an antireflection coating and a sparse contact metal pattern is also added. A simplified cross section of the pixel structure is shown in Figure 2. The APD is a separate-absorber-multiplier structure, and the figure also shows the electric field profile at operational bias. A photon incident from the back side creates an electron-hole pair in the absorber. The hole is collected at the back side, and the electron drifts to the multiplier region, where the avalanche is initiated.

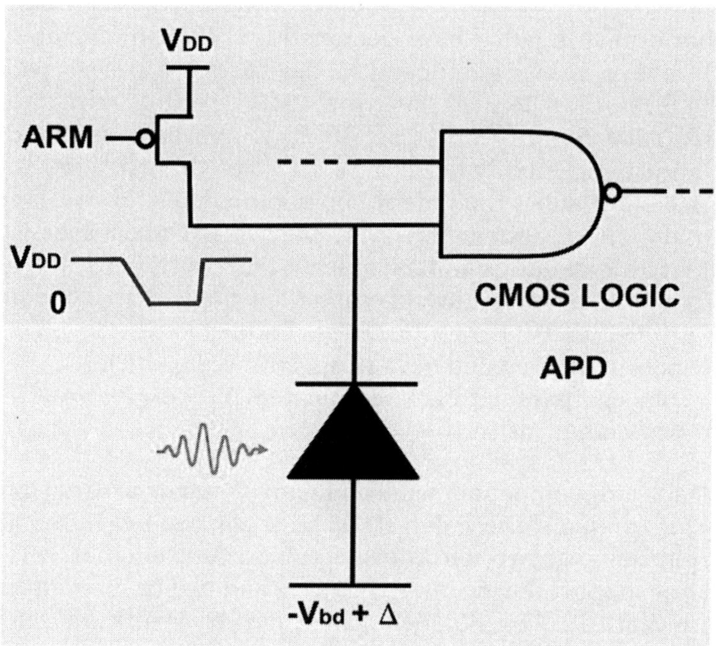

Figure 1. Photon-to-digital conversion. The GMAPD is biased so that it produces a CMOS-logic-compatible voltage pulse, enabling a direct connection to pixel logic.

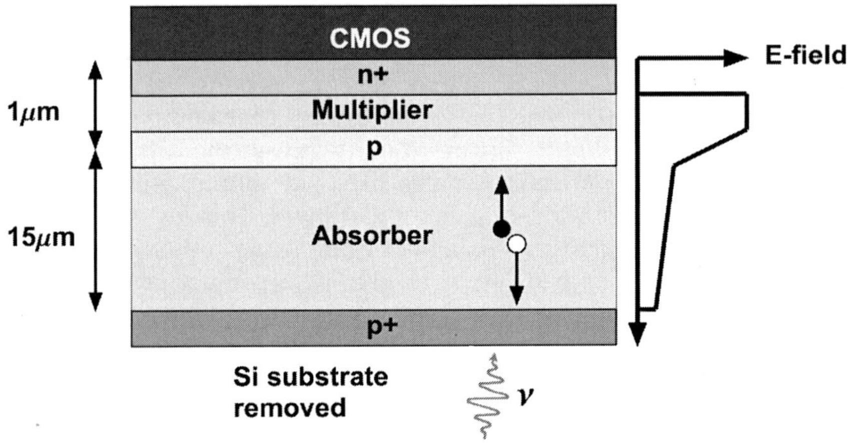

Figure 2. Simplified cross section showing pixel structure, field profile, and operation.

A number of investigators have demonstrated front-illuminated silicon GMAPD arrays using a monolithic device structure in which the photodiode is incorporated into the CMOS readout circuit using a standard CMOS foundry process [10]. This approach facilitates rapid and low-cost prototyping. Because of the thin device structure, monolithic CMOS APDs can time stamp photon detections with very low (tens of ps) timing jitter. However, they share pixel area with the readout circuitry and this limits fill factor. Moreover, fabrication using a standard CMOS process flow precludes customization of photodiode doping profiles and layer thicknesses. Photon detection efficiency is typically poor in the red and near-IR spectral regions. This is the reason why Lincoln has pursued back-illuminated APD arrays hybridized to foundry-fabricated CMOS.

Hybridization of compound semiconductor detector arrays (InGaAs or HgCdTe) is common. These detectors, however, are fabricated in layers heteroepitaxially grown on transparent substrates that serve as a mechanical support during bump bonding and device operation. Silicon APDs are fabricated on homoepitaxially grown layers on a substrate that is opaque at wavelengths of operation. The substrate must be entirely removed either before or after bump bonding, entailing difficulties associated with handling of a thin (15 μm) detector layer or with transfer of this layer to a transparent substrate.

The first GMAPD arrays made at Lincoln were hybridized by a technique known as bridge bonding. Individual CMOS die were epoxied to an APD wafer, providing immediate mechanical support, but with no electrical connection. After removal of the APD wafer substrate, large vias were etched through the detector layer between pixels to expose connection pads, and a metal strap patterned to complete the electrical connection. This technique worked but its use is limited to devices with coarse (>50 μm) pixel pitch and low fill factor [2].

Another technique pursued is transfer and bump bonding [12], shown in Figure 3. Once the APD arrays are fabricated, the wafer is bonded to a temporary silicon handle wafer. The APD substrate is removed and then the device bonded to a transparent fused silica substrate. The temporary handle is removed, and then bump bonding to CMOS readout chips can be carried out in the same manner as for the heteroepitaxial detector materials. Both the temporary handle and the transparent substrate are bonded using oxide-oxide bonding with no adhesive. The transfer process provides APD arrays that can then be quickly bumped to any CMOS readout chip that matches the format and pixel pitch. It is, however, a relatively complex process because of the two-step transfer involved.

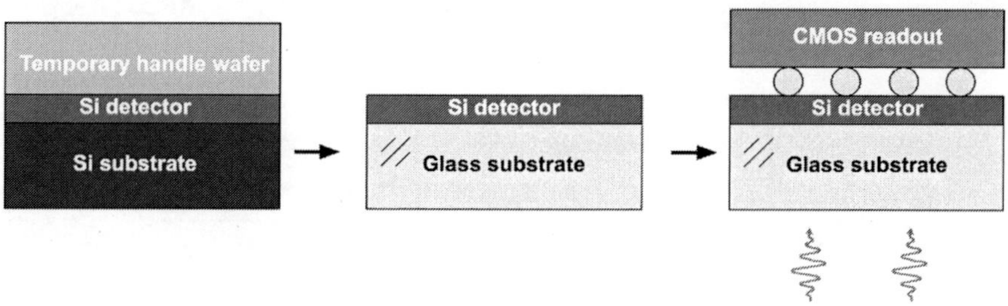

Figure 3. Transfer and bump bonding.

Lincoln Laboratory pioneered a 3D-integration process using its in-house fully depleted silicon-on-insulator (SOI) CMOS process. One of the first imager demonstrations was a 64 × 64-pixel GMAPD array for lidar [13]. The APD was integrated with two tiers of SOI CMOS readout circuitry. Figure 4 illustrates the process. After APD and SOI CMOS fabrication is complete, the APD wafer (tier 1) and the first CMOS wafer (tier 2) are precision aligned and oxide bonded face to face. Then the

SOI handle wafer is removed, using the SOI buried oxide as an etch stop. Concentric vias are then etched through the oxide layers and filled with tungsten to interconnect the last metal of the SOI wafer with the top metal of the APD wafer. These tungsten plugs are micron-scale in diameter and only a few microns in height, much smaller than bump bonds or through-silicon vias used in some wafer stacking processes. The process can be repeated, adding the tier-3 CMOS wafer to the stack. Because the SOI handle wafer is removed, the process is amenable to multiple-tier structures. Figure 5 shows a cross-sectional micrograph of a pixel of the 64 × 64-pixel GMAPD lidar imager.

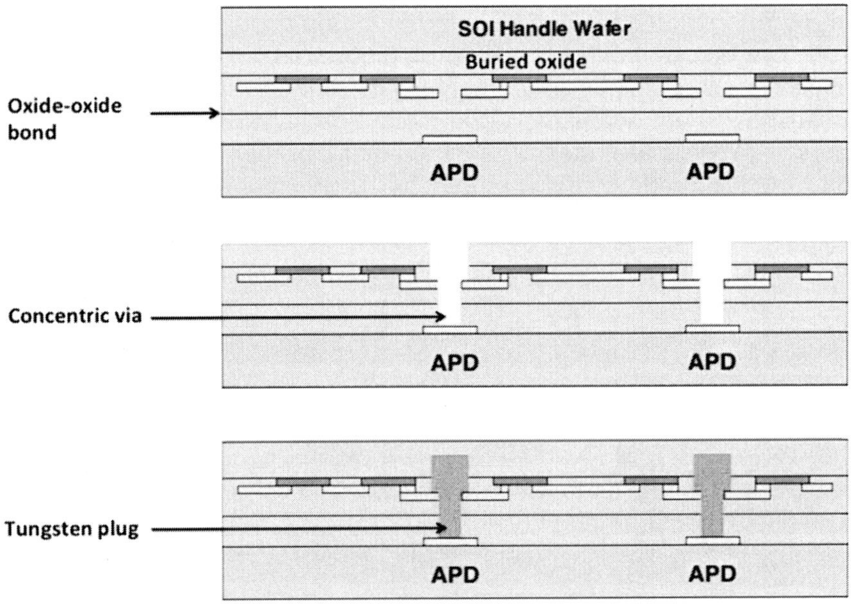

Figure 4. Steps in the Lincoln 3D-integration process to integrate the APD wafer (tier 1) with the first SOI CMOS wafer (tier 2).

3D integration is an important technology direction for advanced image sensors. In the device structure shown in Figure 5, tier 1 contains the photodetectors, tier 2 has first-level signal conditioning circuits, and tier 3 has high-speed logic for photon time stamping. The ability to divide the pixel circuitry among multiple tiers enables a sophisticated pixel circuit to fit into a smaller area than would be possible for a conventional 2D-integrated circuit. A tier could be dedicated to image processing functions to extract information at the pixel level and reduce readout bandwidth. Having multiple tiers also enables the designer to combine multiple circuit or device technologies into a single integrated device, as in Figure 5.

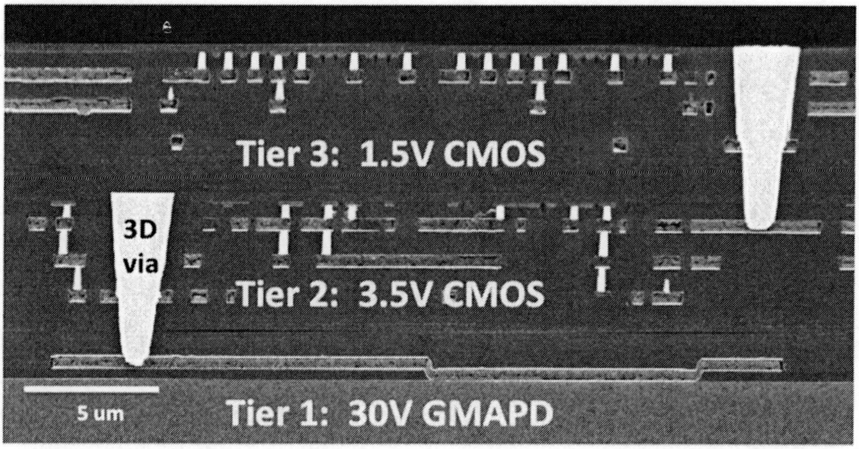

Figure 5. Cross-sectional SEM through a pixel of the 64×64 GMAPD array.

The Lincoln 3D process is an example of a *via-last* process. The connection via from each tier of SOI CMOS down to the previous tier is made after wafer bonding and SOI handle removal. This approach is amenable to multi-tier structures, but requires dedicated full-wafer SOI CMOS for each tier, with the exception of tier 1. One can also use a *via-first* approach [14], illustrated in Figure 6. Metal posts (which perform the same function as bump bonds, but are much smaller) are patterned on each wafer and planarized along with the bonding oxide. When the wafers are bonded and then heated to strengthen the oxide-oxide bond, the posts expand and fuse. In a single step, therefore, mechanical bonding and electrical connection are achieved simultaneously. The CMOS wafer also functions as a permanent mechanical support for APD substrate removal, backside processing, and subsequent imager operation. The *via-first* approach allows for the use of any CMOS process, although it is not easily extendable to multi-tier structures.

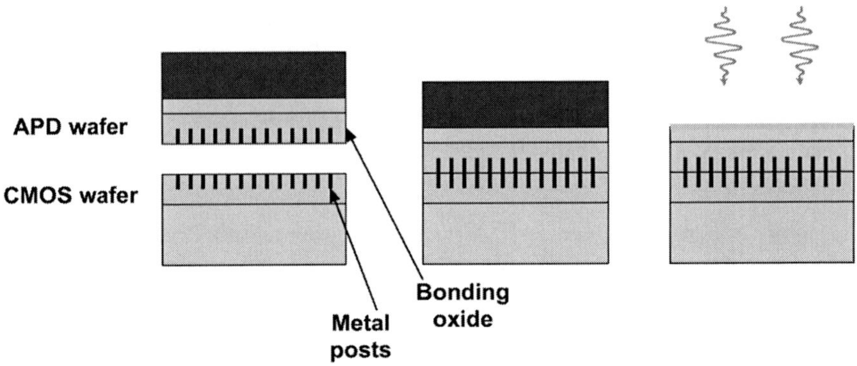

Figure 6. *Via-first* 3D integration.

4. APPLICATION TO LIDAR IMAGING

The first application of Lincoln's GMAPD arrays was optical flash radar for three-dimensional imaging. This is known as ladar or lidar, and we will use the second term. In a flash lidar system the scene is illuminated by a short laser pulse, and imaged onto an array of detectors, each of which measures photon arrival time, and therefore depth to the corresponding point in the scene. By using an array of detectors, one can avoid the mechanical scanning needed in single-detector systems. Because the available light is divided among many pixels, however, the average return signal can be weak, sometimes less than a single photon per pixel. In addition, the pixel circuit is called upon to perform high-temporal-bandwidth time stamping. For linear-mode detectors and analog preamplifiers, sensitivity and speed are conflicting requirements. High noise bandwidth comes along with high signal bandwidth. The user of the system is forced to transmit enough photons so that the amplified return in each pixel can be discriminated from the noise floor. For a given amount of average laser power, this mandates low pulse repetition frequency with high energy per laser pulse.

The use of a Geiger-mode detector array offers a solution. The GMAPD is triggered by a single photon to produce a digital pulse, which is then digitally time stamped by the pixel circuit. There is no tradeoff between sensitivity and timing resolution. The lidar system is operated to take advantage of single-photon time stamping. The laser is operated at high pulse repetition frequency and low energy per laser pulse, so that each pixel gets average returns of a fraction of a photon. The image is built up by combining multiple frames. This mode of operation has two advantages. First, no photons are "wasted" overcoming a noise floor; timing information is obtained from each detected photon. Second, if a pixel sees returns from multiple depths, a histogram of timing values can be built up that reproduces the temporal profile that would be seen by a linear-mode detector operating with a much stronger signal.

The first proof of concept of a GMAPD array for lidar was a front-illuminated 4 × 4-pixel APD die piggy backed onto a 16-channel timing chip and the connections made with wire bonds. This device is shown in Figure 7. The CMOS circuit was fabricated through MOSIS using an HP 0.5-μm foundry process. The timing was done by broadcasting a clock to all the pixels, each of which had a pseudorandom counter based on a linear feedback shift register.

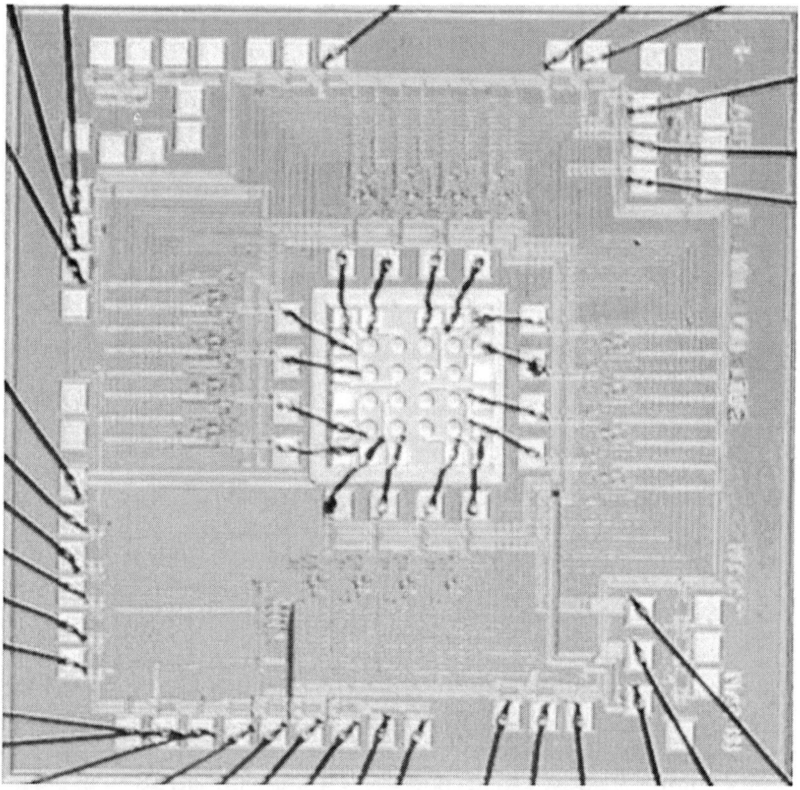

Figure 7. 4 × 4 GMAPD array wire bonded to CMOS timing circuits.

The APD doping profile used in the lidar devices gives a low fill factor. A cross section through two adjacent pixels is shown in Figure 8. The center of the APD has the separate-absorber-multiplier structure depicted in Figure 2, but the n^+-doped region extends out beyond the multiplier, creating a peripheral portion of the junction that collects electrons thermally generated in the region between pixels; these electrons do not trigger Geiger-mode events. A microlens array is integrated on the back side to concentrate incident light onto the responsive portions of the APDs. Alternatively, the lidar transmitter can be designed to project an array of spots onto the scene, which are then imaged on the APDs.

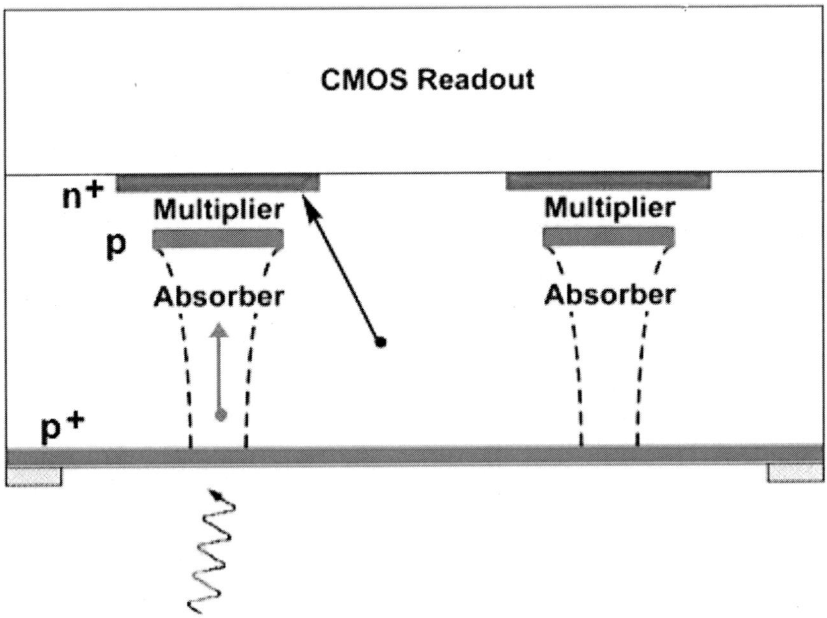

Figure 8. Low-fill-factor APD design used in lidar sensors.

Following this proof of concept, the Laboratory developed a series of 32 × 32 arrays bridge bonded to MOSIS-fabricated CMOS circuits, and used them to build systems to perform foliage penetration and terrain mapping. Figure 9 shows the results of a foliage penetration experiment [15]. Lidar data was collected through a forest canopy from different heights on a tower. These multiple views enable returns to be collected from different angles through gaps in the foliage, thereby filling in the details about objects obscured by the foliage in conventional imagery. Figure 9b is a composite 3D image with the early returns from foliage filtered out, revealing vehicles, picnic tables, and a gazebo.

Figure 9. (**a**) Forest canopy as viewed from a nearby tower (**b**) Composite lidar image obtained by combining data collects from four different heights and then filtering out the early returns.

Lincoln developed the Airborne Ladar Imaging Research Testbed (ALIRT) system, an airborne lidar for terrain mapping. Initially, the system operated at 780 nm and used a 32 × 32 silicon GMAPD array. Because of the availability of efficient lasers at 1060 nm, however, the Laboratory developed short-wave-IR-sensitive GMAPDs based on InGaAsP detectors grown on InP substrates. These APDs also have the separate-absorber-multiplier structure, but the doping profile is created in the epitaxial growth and the pixels are isolated by mesa etch. 128 × 32 lidar image sensors were built by bump bonding the APD arrays to a CMOS timing circuit. Because of the exquisite sensitivity of the GMAPDs, the ALIRT system could collect wide-area terrain maps fifteen times faster than commercially available mapping lidars.

One example of the many missions flown by ALIRT was in support of humanitarian efforts in Haiti after the 2010 earthquake [16]. Figure 10 shows a lidar image of a bridge in Port-au-Prince, Haiti. ALIRT collected terrain maps of the city over time, which enabled relief workers to know where tents were being erected or taken down, what roads were blocked, and where it was safe to land helicopters.

A successor to the ALIRT system has an area coverage rate of 400 km^2/h at 25-cm ground sampling distance. It can rapidly map a broad region and supply detailed three-dimensional images of every terrain feature or manmade structure over which it flies. It sees through foliage or dense dust clouds [17].

Figure 10. Lidar image of a bridge in Port-au-Prince, Haiti.

Lincoln demonstrated another lidar focal plane that is highly significant from a technology evolution standpoint. Figure 11a shows the 3D-integrated 64 × 64-pixel lidar focal plane whose three-tier pixel cross section is shown in Figure 5. Figure 11b is a lidar image of a cone obtained using this device. To our knowledge, this was the first demonstration of a three-tier integrated circuit of any kind, based on a process that supports dense, arbitrarily placed micron-scale inter-tier connection vias (as opposed to chip stacking with peripheral wire bonds or ball grid arrays).

3D integration, as already pointed out, allows one to mix different technologies and put more circuitry within the area of a pixel. More importantly, however, it enables new imager architectures. Raw pixel data flows in parallel up through tiers of on-focal-plane processing circuits that extract information and reduce readout bandwidth.

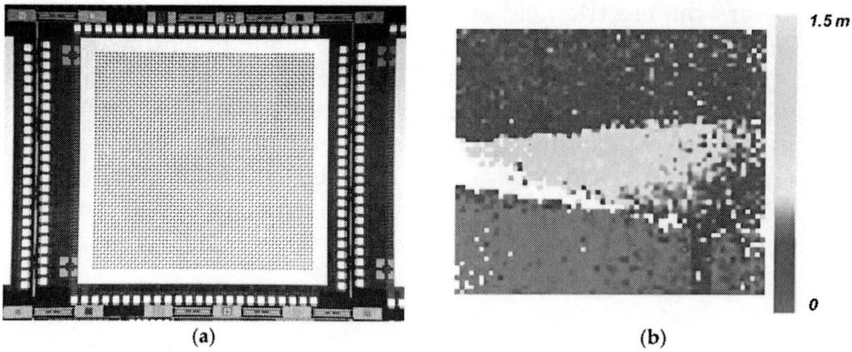

Figure 11. (**a**) 64 × 64-pixel lidar focal plane fabricated by 3D integration of a GMAPD array with two tiers of SOI CMOS circuitry (**b**) Lidar image of a cone obtained by illumination from a doubled Nd:YAG microchip laser.

5. A PHOTON-COUNTING WAVEFRONT SENSOR

Adaptive optics systems for ground-based astronomy and space surveillance require sensors to measure the distortion of a wavefront (from either a bright star or an artificially created beacon) due to atmospheric turbulence. The Shack-Hartmann technique [18] uses arrays of lenslets that focus the light on quad-cell detectors; the displacement of a light spot from the center of a quad cell determines the partition of intensity among the four pixels. This in turn indicates the local wavefront tilt. Lincoln Laboratory has used its CCD imagers to build Shack-Hartmann wavefront sensors. However, in scenarios with weak beacon signals and fast wavefront update rates, the performance of CCDs is limited by readout noise. To address this problem, Lincoln developed Geiger-mode quad-cell arrays, to measure the number of photons from a beacon. Since the quad cell must be responsive to photons incident in between pixels, the low-fill-factor design shown in Figure 8 is not suitable. A high-fill-factor design was devised [19] and its cross section is shown in Figure 12.

The upper p-type layer is implanted at high energy through an oxide mesa so that the doping profile peaks at a relatively shallow depth (1 μm) in the center of each diode and deeper (2 μm) around the periphery and between pixels. The shallow portion of this stepped implant separates the absorber and multiplier portions of each detector. The step lowers the electric field at the periphery of the diode, preventing edge breakdown. The peripheral part of the diode functions as a guard ring, collecting surface-generated dark current without triggering Geiger-mode events. The deep portion of the implant, which is partially undepleted, prevents the guard ring from collecting photoelectrons generated in the absorber; as indicated in the figure, these photoelectrons reach a nearby multiplier region by a combination of diffusion and drift.

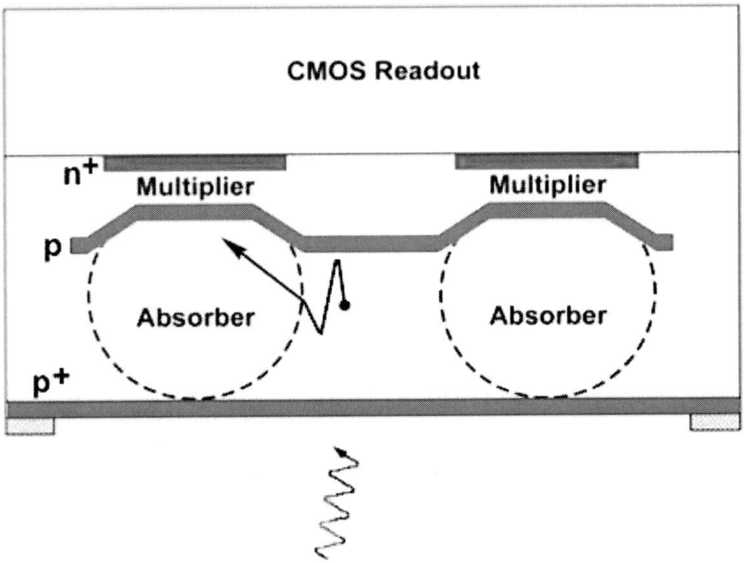

Figure 12. High-fill-factor APD design used in quad cells and passive imaging devices.

16 × 16 and 32 × 32-subaperture quad-cell arrays were fabricated and hybridized to readout chips that count the number of detection events with a 10-bit counter in each pixel. The device could be operated with 20-μs wavefront update latency while introducing no readout noise. Hybridization was accomplished first by bridge bonding and then later on by transfer and bump bonding. Pixel pitch within each quad cell was 50 μm, with a 200-μm spacing between subapertures [20].

To verify the functionality and contiguous spatial response, a focused 5-μm-diameter spot from a blue (450-nm) LED was raster scanned over the area of a quad cell and, at each 5-μm step, recording the photon counts from each of the four pixels.Figure 13a is a contour plot of the count rate of the lower right pixel as a function of the position of the light spot. Figure 13b shows the aggregate count rate from all four pixels. This data shows a monotonic transition of detection activity from one pixel to its neighbor as the light spot is being moved across the midpoint. It also shows no droop in the aggregate response in the central region.

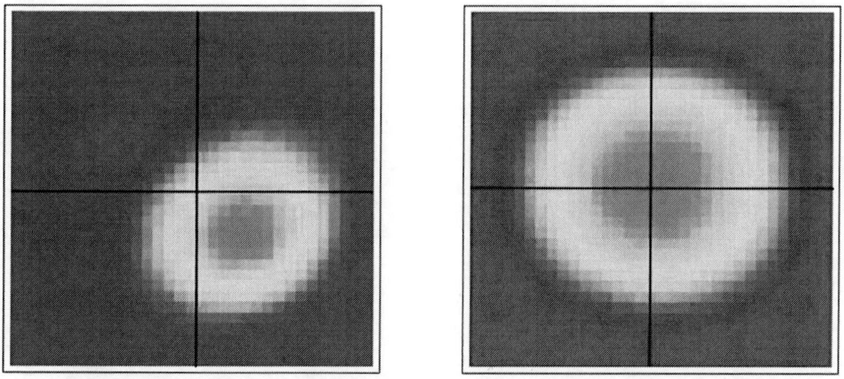

(a) (b)

Figure 13. (**a**) Color contour plot of the count rate from the lower right detector as a function of the position of a small light spot raster scanned over the area of the quad cell; (**b**) Contour plot of the sum of the count rates from all four pixels as a function of light spot position.

These quad-cell arrays report out raw pixel intensity values, and the wavefront tilt calculation is done off chip. However, one can envision incorporating computational functions such as centroid and tracking into the pixel circuitry.

6. PASSIVE IMAGING

In 2011, the Laboratory demonstrated a 256 × 256 passive photon counting imager with 25-μm pixel pitch [21]. The GMAPD array was based on the high-fill-factor design (see Figure 12) and was integrated to a foundry-fabricated CMOS readout using the transfer and bump bonding technique. To our knowledge, this was the first ever back-illuminated passive image sensor with this large a format based on hybridization of a GMAPD array to a CMOS readout. Figure 14 shows one of the first images taken with the device, a church steeple located about 3.5 km from the camera system. The black specks in the image are bump bond defects. Use of a 3D-integration technique results in far better image cosmetics.

Figure 14. Image of church steeple.

The GMAPDs are armed, queried, and reset under supervision of external polling clocks. The CMOS pixel circuit has two readout modes. On the one hand, one can read out binary images in which each pixel reports whether or not it had a detection event since the last readout. This readout mode is called binary readout mode. On the other hand, one can read out the overflow bit of a 7-bit pseudorandom counter in each pixel; if set, this bit represents 127 cumulative detection events and it is reset by the readout operation. This readout mode is called overflow readout mode. At the conclusion of multiple overflow-bit readouts, the entire 7-bit counter can be read out to get the remainder. Binary readout mode and overflow readout mode are implemented by separate clocking systems that address rows or columns, respectively, in a rolling readout that does not blind the detectors. The overflow-bit readout mode provides for readout bandwidth reduction and dynamic range extension without having to put a large counter in the pixel. The overflow bits can be streamed to an Field-Programmable Gate Array (FPGA) that effectively provides the most significant bits of a longer counter. The image in Figure 14 was obtained by operating a 8 kiloframes/s in binary readout mode, digitally adding many frames off chip and performing a flat field correction.

Figure 15 shows images resulting from post-readout digital summation of short-time binary frames. Since there is no readout noise in a photon counting device, there is no noise penalty for such post-readout summation. One can imagine processing the binary frames before summing them to correct for scene motion or platform vibration. Even if each binary frame has a signal level less than one photoelectron per pixel, enough pixels have events to create discernable spatial structure, making such "smart integration" possible. This would not be feasible with a conventional analog imager, as even 1 electron rms readout noise would be too much.

A technological challenge with dense-pitch Geiger-mode APD arrays is optical crosstalk. During a detection event, the carriers traversing the avalanche region lose some energy by optical emission, and the near-infrared photons given off can spuriously trigger neighboring pixels. The quad-cell arrays [22] and the passive imager yielded valuable data on this phenomenon. Above a certain bias voltage, crosstalk-induced events dominate the dark count rate. Optical-crosstalk-based analytical models and Monte Carlo simulations match the spatial, temporal, and statistical characteristics of the dark count activity observed in the image data. The APDs in the passive imager did not have aggressive crosstalk reduction features, because that would have added yet another element of yield risk in the effort to prove out a new hybridization technique. The devices are operated at sufficiently low bias to avoid the crosstalk-dominated dark-count-rate regime. This limits the photon detection efficiency to the 10%–20% range. With the maturation of 3D integration techniques, current efforts are focused on crosstalk reduction.

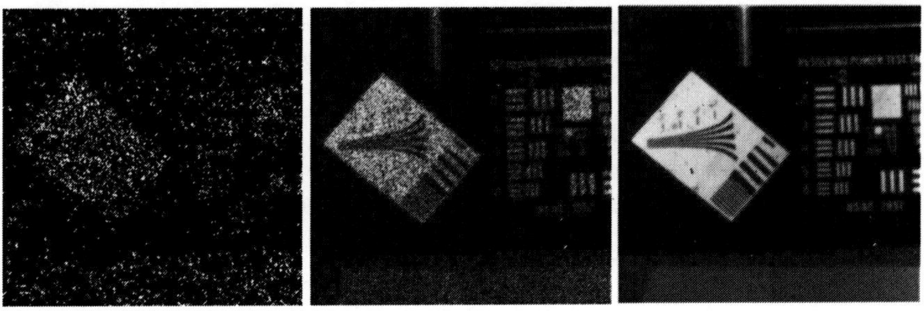

(a) (b) (c)

Figure 15. (**a**) One binary frame; (**b**) the digital sum of 64 binary frames; and (**c**) the digital sum of 32,768 binary frames.

7. THE FUTURE OF GMAPD IMAGER TECHNOLOGY

The past 20 years have seen great progress in solid-state image sensors based on custom Geiger-mode avalanche photodiode arrays hybridized or 3D-integrated to all-digital CMOS readouts. Application to lidar represented the "low-hanging fruit," as compelling system functionality could be realized with small imager format, coarse pixel pitch, low fill factor, and dark count rates in the tens of kcounts/s. Passive imaging is far more demanding of APD performance and, for back-illuminated silicon arrays, much more dependent on the maturation of hybridization techniques. Now that 3D integration methods have become available, optical crosstalk reduction is the next task. The successful demonstration of a 256 × 256 passive photon counting imager without crosstalk reduction is an encouraging result. A number of measures, such as capacitance scaling and low-reflectivity contact metal, can now be implemented to improve sensitivity.

All-digital CMOS tiers can exploit Moore's Law scaling to realize increasingly sophisticated on-focal-plane processing functions. For lidar, these include in-pixel histogramming, tracking of objects of interest, multi-frame coincidence to reject background, and data thinning. For passive imaging, these include image stabilization, temporal change detection, and spatial filtering. Ultimately, one could implement a deeply scaled CMOS tier that could be programmed like an FPGA and support multiple firmware-defined functions with no hardware redesign.

While the focus of this review is imaging, GMAPD arrays sensitive in the short-wave infrared have potential as agile laser communications receivers. By spreading the received laser flux over multiple pixels, the effective APD reset time can be much shorter than the single-detector reset time, enabling high-data-rate free space optical communications links. In a remarkable technology demonstration, researchers at the Laboratory demonstrated the feasibility of an optical data link from a science satellite orbiting Mars. The laboratory demonstration used a pulse-position modulated format and error correction coding to achieve 0.5 photons/bit [23].

An imaging system traditionally consists of three distinct subsystems: (1) A bulky optical train that merely carries out an isomorphic transformation from object space to image space; (2) an image sensor that produces a high-aggregate-bandwidth stream of raw image data; and (3) a post-processing system to extract information of interest. Lincoln Laboratory's long-term vision is to merge these functions, so that the work of information extraction is carried out by co-designed computational optics and smart focal planes. The use of coded apertures or light-field camera architectures can be combined with digital time stamping of photon arrivals to extract information about a scene, including regions that are obscured from the view of a conventional camera. The shift of computation burden to the optics and imager could also reduce readout bandwidth and enable low size, weight, and power.

CONFLICTS OF INTEREST

The authors declare no conflict of interest.

REFERENCES

1. Burke, B.E.; Gregory, J.A.; Cooper, M.; Loomis, A.H.; Young, D.J.; Lind, T.A.; Doherty, P.; Daniels, P.; Landers, D.J.; Ciampi, J.; *et al.* CCD Imager Development for Astronomy. *Linc. Lab. J.* **2007**, *16*, 393–412.

2. Aull, B.F.; Loomis, A.H.; Young, D.J.; Heinrichs, R.M.; Felton, B.J.; Daniels, P.J.; Landers, D.J. Geiger-Mode Avalanche Photodiodes for Three-Dimensional Imaging. *Linc. Lab. J.* **2002**, *13*, 335–350.

3. McIntosh, K.A.; Donnelly, J.P.; Oakley, D.C.; Napoleone, A.; Calawa, S.D.; Mahoney, L.J.; Molvar, K.M.; Duerr, E.K.; Shaver, D.C. Development of Geiger-mode APD arrays for 1.06 μm. In Proceedings of the 15th Annual Meeting of the IEEE Lasers and Electro-Optics Society, Glasgow, Scotland, 10–14 November 2002; pp. 760–761.

4. Duerr, E.K.; Manfra, M.J.; Diagne, M.A.; Bailey, R.J.; Zayhowski, J.J.; Donnelly, J.P.; Connors, M.K.; Grzesik, M.J.; Turner, G.W. Antimonide-based Geiger-mode avalanche photodiodes for SWIR and MWIR photon-counting. *Proc. SPIE* **2010**, *7681*.

5. Frechette, J.P.; Grossmann, P.J.; Busacker, D.E.; Jordy, G.J.; Duerr, E.K.; McIntosh, K.A.; Oakley, D.C.; Bailey, R.J.; Ruff, A.C.; Brattain, M.A.; *et al.* Readout circuitry for continuous high-rate photon detection with arrays of InP Geiger-mode avalanche photodiodes. *Proc. SPIE* **2012**, *8375*.

6. Gersbach, M.; Trimananda, R.; Maruyama, Y.; Fishburn, M.; Stoppa, D.; Richardson, J.; Walker, R.; Henderson, R.K.; Charbon, E. High Frame-rate TCSPC-FLIM Using a Novel SPAD-based Image Sensor. *Proc. SPIE* **2010**, *7780*.

7. Gordon, K.J.; Hiskett, P.A.; Lamb, R.A. Advanced 3D Imaging Lidar Concepts for Long Range Sensing. *Proc. SPIE***2014**, *9114*.

8. Grein, M.E.; Elgin, L.E.; Robinson, B.S.; Caplan, D.O.; Stevens, M.L.; Hamilton, S.A.; Boroson, D.M.; Langrock, C.; Fejer, M.M. Efficient communication at telecom wavelengths using wavelength conversion and silicon photon-counting detectors. *Proc. SPIE* **2007**, *6709*.

9. Hadfield, R. Single-photon detectors for optical quantum information applications. *Nat. Photonics* **2009**, *3*, 696–705.

10. Charbon, E. Towards large scale CMOS single-photon detector arrays for lab-on-chip application. *J. Phys. D Appl. Phys.* **2008**, *41*.

11. Cova, S.; Ghioni, M.; Lacaita, A.; Samori, C.; Zappa, F. Avalanche photodiodes and quenching circuits for single-photon detection. *Appl. Opt.* **1996**, *35*, 1956–1976.

12. Schuette, D.R.; Westhoff, R.C.; Loomis, A.H.; Young, D.J.; Ciampi, J.S.; Aull, B.F.; Reich, R.K. Hybridization process for back-illuminated

silicon Geiger-mode avalanche photodiode arrays. *Proc. SPIE* **2010**, *7681*.

13. Aull, B.F.; Burns, J.; Chen, C.; Felton, B.; Hanson, H.; Keast, C.; Knecht, J.; Loomis, A.; Renzi, M.; Soares, A.; *et al.* Laser radar imager based on three-dimensional integration of Geiger-mode avalanche photodiodes with two SOI timing-circuit layers. In Proceedings of the 2006 International Solid State Circuits Conference Digest of Technical Papers, San Francisco, CA, USA, 6–9 February 2006; pp. 304–305.

14. Enquist, P.; Fountain, G.; Petteway, C.; Hollingsworth, A.; Grady, H. Low Cost of Ownership Scalable Copper Direct Bond Interconnect 3D IC Technology for Three Dimensional Integrated Circuit Applications. In Proceedings of the IEEE International Conference on 3D System Integration, San Francisco, CA, USA, 28–30 September 2009.

15. Aull, B.F. 3D Imaging with Geiger-mode Avalanche Photodiodes. *Opt. Photonics News* **2005**, *16*, 42–46.

16. Neuenschwander, A.L.; Crawford, M.M.; Magruder, L.A.; Weed, C.A.; Cannat, R.; Fried, D.; Knowlton, R.; Heinrichs, R. Terrain classification of ladar data over Haitian urban environments using a lower envelope follower and adaptive gradient operator. *Proc. SPIE* **2010**, *7684*.

17. The Discussion of this in the Context of Lincoln Laboratories Effort in Humanitarian Assistance and Disaster Relief. Available online: http://phys.org/news/2015–10-technology-disasters.html (accessed on 5 April 2016).

18. Platt, B.C.; Shack, R. History and principles of Shack-Hartmann wavefront sensing. *J. Refract. Surg.* **2001**, *17*, S573–S577.

19. Renzi, M.J.; Aull, B.F.; Reich, R.K.; Kosicki, B.B. High-fill-factor avalanche photodiode. U.S. Patent No. 8,093,624, 10 January 2012.

20. Aull, B.F.; Renzi, M.J.; Loomis, A.H.; Young, D.J.; Felton, B.J.; Lind, T.A.; Craig, D.M.; Johnson, R.L. Geiger-Mode Quad-Cell Array for Adaptive Optics. In Proceedings of the Conference on Lasers and Electrooptics, San Jose, CA, USA, 4–9 May 2008.

21. Aull, B.F.; Schuette, D.R.; Young, D.J.; Craig, D.M.; Felton, B.J.; Warner, K. A Study of Crosstalk in a 256 × 256 Photon Counting Imager Based on Silicon Geiger-Mode Avalanche Photodiodes. *IEEE Sens. J.* **2015**, *15*, 2123–2132.

22. Aull, B.F.; Reich, R.K.; Ward, C.M.; Craig, D.M.; Young, D.J.; Johnson, R.L. Detection statistics in Geiger-mode avalanche photodiode quad-cell arrays with crosstalk and dead time. *IEEE Sens. J.* **2015**, *15*, 2133–2143.

23. Mendenhall, J.A.; Candell, L.M.; Hopman, P.I.; Zogbi, G.; Boroson, D.M.; Caplan, D.O.; Digenis, G.J.; Hearn, D.R.; Shoup, R.C. Design of an Optical Photon Counting Array Receiver System for Deep-Space Communications. *IEEE Proc.* 2007, *95*, 2059–2069.

2

SELF-BALANCING POSITION-SENSITIVE DETECTOR (SBPSD)

Ryan Porrazzo *, Leigh Lydecker, Suhasini Gattu, Hassaram Bakhru, Natalya Tokranova and James Castracane

Colleges of Nanoscale Science and Engineering, SUNY Polytechnic Institute, 257 Fuller Road, Albany, NY 12222, USA

ABSTRACT

Optical position-sensitive detectors (PSDs) are a non-contact method of tracking the location of a light spot. Silicon-based versions of such sensors are fabricated with standard CMOS technology, are inexpensive and provide a real-time, analog signal output corresponding to the position of the light spot. An innovative type of optical position sensor was developed using two back-to-back connected photodiodes. These so called self-balancing position-sensitive detectors (SBPSDs) eliminate the need for external readout circuitry entirely. Fabricated prototype devices demonstrate linear, symmetric coordinate characteristics and a spatial resolution of 200 µm for a 74 mm device. PSDs are commercially available only up to a length of 37 mm. Prototype devices were fabricated with various lengths up to 100 mm and can be scaled down to any size below that.

KEYWORDS

PSD; back-to-back connected photodiode; linear position sensor; optical position sensor

1. INTRODUCTION

Position-sensitive detectors are used to track the position of a light spot in one or two dimensions [1]. Some common applications of PSDs is range determination by triangulation using a laser and obstacle detection [2,3]. Only 1D PSDs will be discussed in this work. There are two main classes of optical position sensors, or position-sensitive detectors (PSDs). The first type is composed of discrete photo sensor elements that provide position information in the form of a digital output (e.g., CCDs [4] or linear array of discrete photodiodes). The second type is a sensor with a continuous photosensitive surface that provides position data in the form of an analog output (e.g., Lateral effect PSD [5,6]).

The lateral photo effect was officially discovered by Wallmark in 1957, but the effect was originally described by Schottky in 1930 [1,7]. The effect was further developed by Lucovsky in 1960, and has since been developed by other research groups [8,9,10]. The basic structure of a lateral effect PSD is a line composed of a through-wafer photodiode with a contact on either side of the top of the line, and a common contact along the backside of the device. The device is reverse-biased in order to increase the width of the depletion region of the photodiode, and thus the sensitivity to light. Carriers are generated by incident light in the depletion region of the photodiode, and drift through the bulk of the device due to a junction field induced by the applied bias. The generated current is then divided between the two top contacts as a function of the resistance between the incident light spot and each respective contact. Using the ratio of these two currents, the position sensitive detector reveals the median of the incident light spot that is independent of the size or shape of the light spot within the active area.

Improvement of thin film technology has allowed for the creation of large area 1D thin film PSDs along with the ability to create nanostructured silicon for applications in solar cells, position sensors, and thin film transistors [11,12].

Lateral effect position-sensitive detectors (LEPSDs) provide an analog output, offer high resolution, and are a relatively low cost, accurate method of sensing position using standard CMOS processing techniques [13,14]. However, packaging a sensor that needs a backside contact requires a higher packaging cost. If a sensor could have all contacts on one side of the substrate, then the cost of the sensors and their packaging could be reduced by eliminating all backside processing and associated costs. Lastly, minimization of required readout circuitry is desired. Analog output position sensitive detectors typically require less signal processing circuitry than a digital output position sensor. However, developing a sensor with a direct readout from the sensor itself eliminates the need for any external circuitry, which makes using the device for various applications easier and less expensive. Consideration of these factors led to the development of a new type of sensor that will be described in this work.

1.1 Device Operation

The self-balancing position-sensitive detector (SBPSD) consists of two parallel doped lines in a substrate of the opposite doping type with a metal line connecting the contact pads on top of both devices. The line where voltage is applied shall be known as the distributed bias line while the line which the final potential is measured is called the floating line. In our case, two n-type lines were produced in a p-type substrate. Each line is resistive, based on the doping concentration used. The two lines form a linear back-to-back photodiode pair from the n-p-n junction along the device, resulting in the characteristic I-V curve shown in Figure 1. Because the photodiodes are connected back-to-back, one of them is reverse-biased and the other is forward-biased. The reverse bias determines the current flowing inside the device.

The floating line must be measured with a high input resistance device in order to maintain current flow inside the device [15]. Once illuminated, one of the diodes will be split into a part that is reverse-biased on the right side of the illuminated spot, while the left side will be forward-biased. The second diode will then be forward-biased on the right side of the illuminated spot while the left side will be reverse-biased. The potential difference across the n-p-n junction determines the state on the diode, which part of the diode is reverse-biased, which part of the diode is forward-biased, and the resulting direction of current flow. When a light spot is incident on the pair of back-to-back connected photodiodes, the reverse-biased part of the diode generates current. This current travels freely through the forward-biased part of diode, accumulating charge on the floating line and thus changing its potential [16]. The biased line maintains a constant linear potential distribution.

Potential difference between lines at each point across the device determines current direction at each location. The illuminated light spot creates a flow of current between the lines as shown in Figure 2. The generated currents automatically balance their flows around the median of the light spot. The equipotential is the position where the potential values of two lines are the same. At equilibrium the median of the light spot and the device equipotential are located at the same position, allowing for determination of the spot position relative to the device. The flow of current between the two lines causes the potential on the floating line to equilibrate to the same value as that of the biased line at the median of the light spot. At the equilibrium, the currents on both side of the center of the light spots are equal [17].

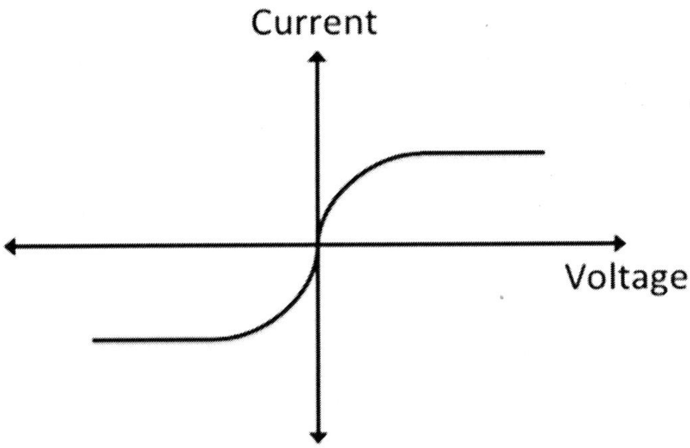

Figure 1. I-V curve of n-p-n junction.

A critical figure of merit for photodiodes is their responsivity, which is the ratio of the optical generated current to the input power of the light, as shown in Equation (1) [18].

$$\Re = \frac{I_{op}}{P_{opt}} \tag{1}$$

The responsivity is given in units of amps per watt (A/W). Typical responsivity values of silicon photodiodes at 623 nm is ~0.4 A/W [19].

In order to determine the responsivity of the position sensitive detectors presented, I-V curves were measured at various LED intensities controlled by the applied current as well as with no illumination. The photogenerated current was obtained by subtracting the measured dark current from the illuminated response. The light intensity was measured using calibrated photodiodes inside a Newport 1815-C Power Meter. The spot used for measurement was completely inside the photosensitive area of the device, the same as in the calibrated photodiode. The responsivity of the prototype detectors in this work was determined to be 0.43 A/W after averaging measurements on three devices, as shown in Figure 3, which compares favorably with commercially available photodiodes illuminated with a 623 nm wavelength LED.

2. EXPERIMENTAL SECTION

6″ (100) p-type silicon wafers with a resistivity of 80–130 ohm-cm, a thickness of 675 μm, and 500 nm of oxide on the surface were used for the device fabrication. First, alignment marks were patterned onto the front of the substrate using standard contact photolithography processing. The alignment marks were etched through the SiO_2 into the silicon wafer itself, providing a permanent reference feature for future alignment.

The main doped device lines were then defined onto the SiO_2 with standard contact photolithography processing, leaving only the desired doping area exposed. The parallel device lines were then doped with phosphorous through the oxide using ion implantation. The implant energy was 350 keV and the dose was 1.67 × 1016 ions/cm^2, as shown in Figure 4, simulated by SRIM (Stopping Range of Ions in Matter) software. After the implant, the photoresist mask was stripped, and the wafers were annealed in argon at 1000 °C for 1 min using a rapid thermal anneal process to thermally activate dopants and fix damage in the crystal lattice.

Openings in the SiO_2 were patterned using contact photolithography and etched using buffered oxide etchant (Transene Company, Inc. Danvers, MA, USA) in order to allow direct contact to each of the doped silicon lines. The electrode metals, 200 Å of titanium followed by 1500 Å of gold, were then deposited onto the substrate, and subsequently patterned to form the electrical device contacts. Upon completion, the devices were sent for dicing.

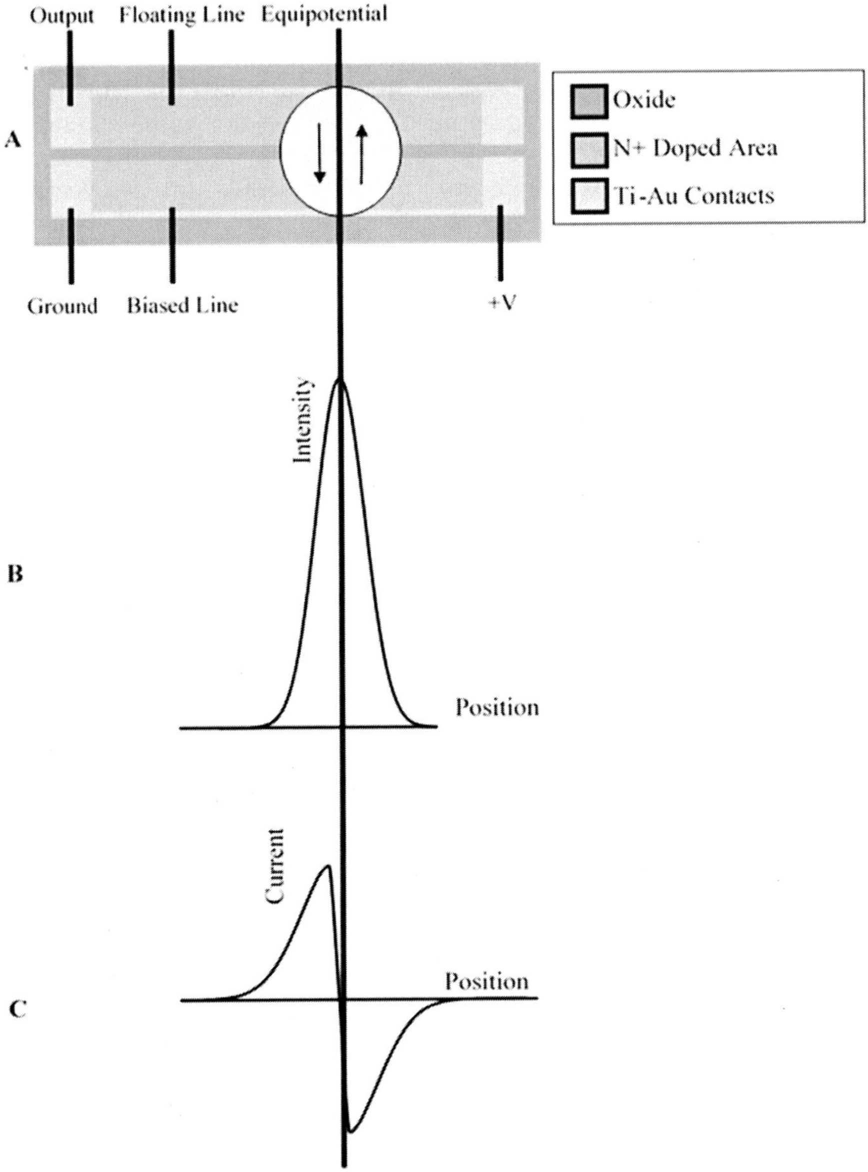

Figure 2. (**A**) Illustration of direction of current flow under illuminated light spot at the equilibrium position; (**B**) spatial distribution of the intensity of the light spot; (**C**) current produced in the device upon illumination in equilibrated state.

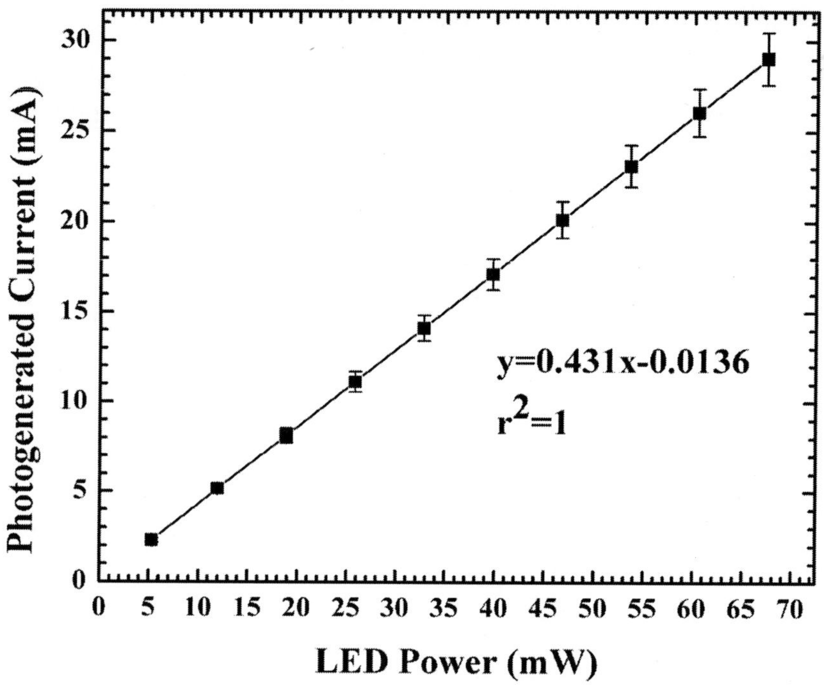

Figure 3. Photogenerated current from SBPSD as a function of incident LED power, showing the device responsivity in A/W as the slope of the curve.

In order to analyze the influence that the width of the separation line between the doped areas has on device performance, devices were fabricated at lengths of 37 mm, 74 mm and 100 mm.

Measurements were performed directly on the device with no external electronic circuitry. Device characterization was done by incrementally moving a light spot along the device using a LabVIEW program that moved a microscope head attached to a Cascade Microtech Alessi REL-6100 programmable probe station to incrementally scan the length of the device with the light spot focused by a microscope. A Keithley 238 source measure unit (SMU) was used to read and record the output voltage while a second Keithley 238 SMU was used to apply the 0 to 5 V bias. Before measurements were performed, the light spot was first centered to fit the entire device width. However, if the device were misaligned or the light spot did not cover the whole device, it would cause asymmetrical illumination on the device, which would affect the resulting coordinate plot. The response of the device to a moving light spot is illustrated by the coordinate plots to follow in the paper. The coordinate characteristic of the PSD is a plot of light spot position *vs.* the output voltage. Self-balancing position sensors were fabricated, packaged and tested.

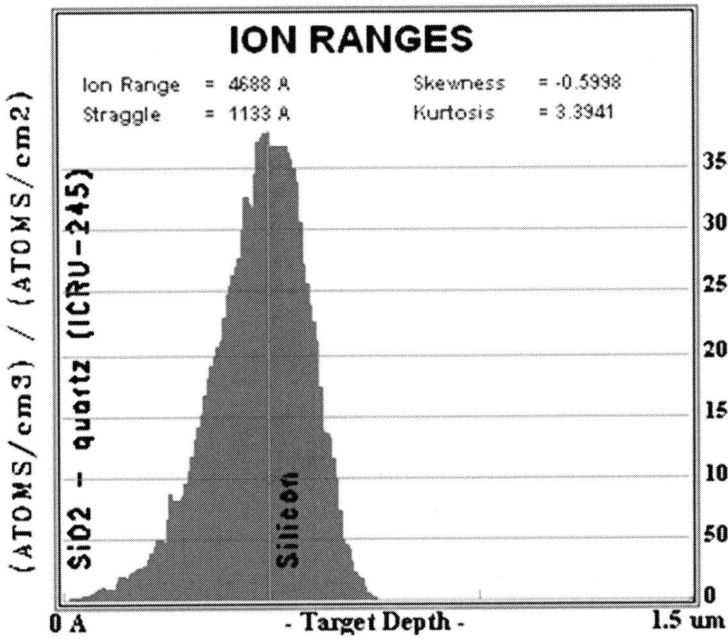

Figure 4. Simulation of hosphorous ion implantation through surface oxide using SRIM software.

3. RESULTS AND DISCUSSION

To investigate this effect, 37 mm-long devices fabricated with a metal line connecting the contact pads for doped lines while bare silicon connected the other doped lines were used for testing (Figure 5). Testing was performed by moving the light spot incrementally by an offset in the vertical direction by 0.2 mm from the center, and the same measurement was performed. The light spot was further offset 0.2 mm to the left for a total of 0.4 mm from the center light spot and a coordinate plot was measured (Figure 5). The same measurement was then performed on the bottom portion of the device at 0.2 mm and 0.4 mm from the center. In Figure 6, points E and F are positioned identically with reference to the light spot, with the only difference being which line the 0 to 5 V bias was applied to. Point E clearly demonstrates that application of a bias onto the metal line achieved a linear profile compared to the curved profile in point F, where the bias was placed on the side with no metal line connecting the contacts. A metal line connecting the contact pads across the device led to fewer losses when compared to a nonmetal line. Lateral displacement in the light spot from the nonmetal to the metal causes the resistance in the nonmetal to decrease, leading to a more linear coordinate plot.

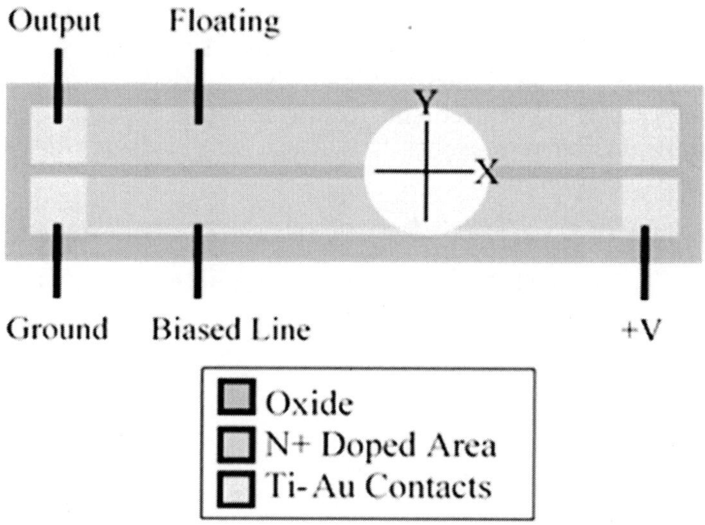

Figure 5. Schematic drawing of light spot position.

I-V measurements were performed to further understand changes in the coordinate characteristic due to variation in spot location. Measurements were performed using a light spot with a Gaussian profile which led to a non-symmetric volt-ampere characteristic for these devices with respect to the position of the light spot across the device. I-V curves were measured by sweeping from −2 to 2 V and recording the corresponding current. If the light spot does not illuminate part of both lines identically, the I-V curve changes in response. This effect was confirmed by measuring the 37 mm-long

SBPSD with 50 micron spacing between the doped lines by first placing the light spot such that it covered the whole device width and measuring the IV response.

The location of the light spot was changed similarly to Figure 5, by moving the spot position by 0.2 mm and 0.4 mm in the Y and −Y direction and an I-V curve was measured at each spot location. In the resulting I-V curves (Figure 7), the positive saturation current decreased for all measurements that were not in the center of the device. For the negative saturation current, a noticeable difference was observed; depending on which side of the device the light spot was moved, the current changed in response. When the light spot was offset towards the biased line, the saturation current was more negative (lower) than at the center light position. When the light spot was offset towards the floating line the current was less negative (higher) than the center light position. This change in the I-V response led to a corresponding change in the coordinate characteristic for these devices. The performance measurements confirmed the importance of light positioning on the photosensitive surface of the device for accurate coordinate determination.

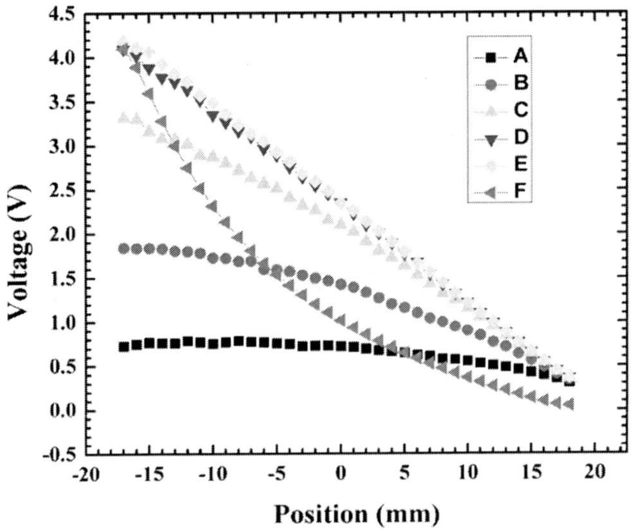

Figure 6. Output voltage *vs.* position measurements for various light positions (A = −0.4 mm, B = −0.2 mm, C = 0 mm, D = 0.2 mm, E = 0.4 mm, F = 0.4 mm). At positions A–E the 0–5 V bias was applied to the bias line and the floating potential measured on the floating line. For position F, the 0–5 V bias was applied on the floating line and the floating potential measured on the biased line.

Investigations were performed to determine how the distance between doped lines would affect the slope of the I-V curve. 37 mm devices were fabricated with a distance of 50 μm and 75 μm between the doped lines. Various distances between the doped lines will result in different values in resistance between lines and changes in the slope of the transition part between saturation regions of I-V curves as can be seen in the inset in Figure 8.

The distributed bias line is the doped silicon and its resistance can be changed by light illumination. Results of the above investigation regarding changes in the coordinate plot based on light spot displacement led to the development of a PSD with dual metal lines as shown in Figure 2a. The addition of dual metal lines eliminated the effect of lateral displacement in the light spot. The influence of resistance change due to photosensitivity on SBPSD coordinate response was studied. The coordinate characteristic of the device and its resistance along the biased line were measured to determine the effect of resistance change on the linearity in the coordinate plot. Two types of devices were used for this measurement. The devices that were used for this measurement are the dual metal line 100 mm length devices. The metal used for the contact pads is 1500 Å of Au on top of 500 Å of Ti. In one device the contact pads were covered with chemically resistant tape and the Au was etched, leaving only Ti on the line connecting the device. The device with only Ti across the device had a maximum resistance of 26.8 kΩ near the side where 5 V was applied compared to the device with Au-Ti, yielding a maximum resistance of 163.7 Ω.

A 0-to-5 volt bias was first applied across the Ti line. Using this bias a coordinate plot for voltage and resistance change *vs.*position were measured, and the resistance was calculated by Ohm's law. A significant resistance change (~70%) led to a non-linear, non-symmetric coordinate plot (R^2 = 0.9387) (Figure 9).

The 0-to-5 V bias was then applied to the Au-Ti line. Using this bias condition, a coordinate plot for voltage and resistance change *vs.* position was measured. It was observed that having a low resistance led to a linear symmetric coordinate plot. The resistance was essentially constant across the device with an overall resistance change of 0.2%. The application of the bias to the device with the lower resistance led to a linear symmetric coordinate plot and a significantly lower resistance change as seen in Figure 10. For greater device lengths (>37 mm), dual metal lines are needed to achieve a linear coordinate plot due to the large non-uniformity in the potential distribution across the floating line that occurs when there is no metal line connecting the contacts.

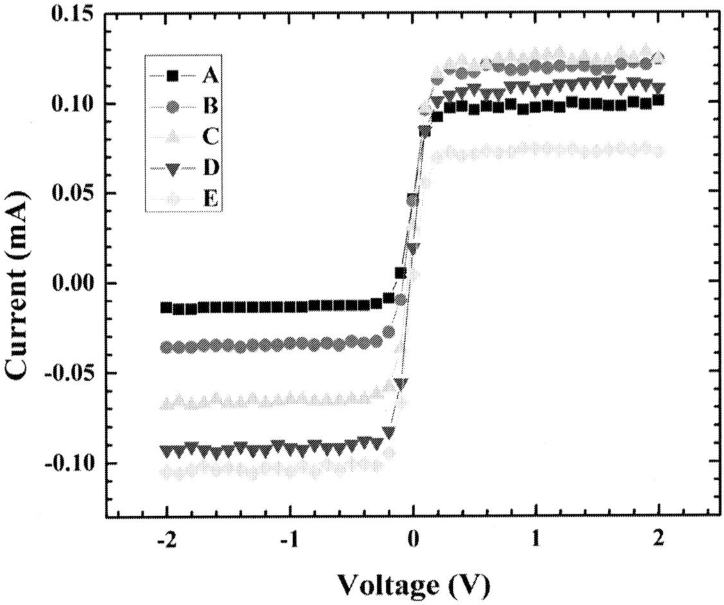

Figure 7. Output voltage *vs.* current measurements for various light positions (A = −0.4 mm, B = −0.2 mm, C = 0 mm, D = 0.2 mm, E = 0.4 mm).

The theoretical spatial resolution of a PSD is the minimum detectable light spot displacement. For a 1-D PSD, as the active area increases, the position resolution decreases. For an active area of 6 mm, 12 mm and 37 mm, the corresponding position resolution is 0.2 μm, 0.3 μm and 2.8 μm respectively [20]. The theoretically smallest detectable position displacement can be expressed as

$$\Delta x = L \times \frac{I_n}{I_{Tot}} \tag{2}$$

where L is the length of the detector, I_{Tot} is the total photocurrent from all contacts, and I_n is the noise current [13]. The theoretical spatial resolution for a 74 mm device was determined to be approximately 200 μm using Equation (2) in conditions where the photogenerated current exceeds the dark current by approximately 300 times. Theoretical spatial resolution was then verified by moving the microscope head on the probe station in 200-μm steps and the output voltages were recorded (Figure 11). A plot of voltage *vs.* position shows a highly linear coordinate plot ($R^2 = 0.99529$) over a 4-mm distance.

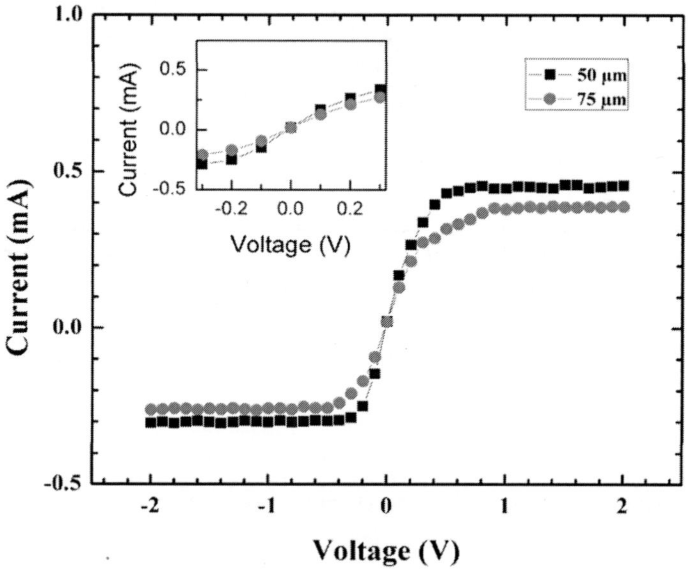

Figure 8. I-V curve of 37 mm SBSPD with 50 μm and 75 μm spacing between doped lines. Inset from −0.3 to 0.3 shows the resistance changes when varying the distance between doped lines.

A symmetric and linear coordinate plot is necessary for operation of the self-balancing position-sensitive Detector. The sensor's operation was tested using Microsoft Windows built-in game controller configuration software. This configuration software allows for the user to set a voltage value inside a defined range to a light spot position along a particular axis. In this case, the output voltage varies with the movement of the LED and is defined as movement on the x-axis. The operation of the sensor is demonstrated in Figure 12.

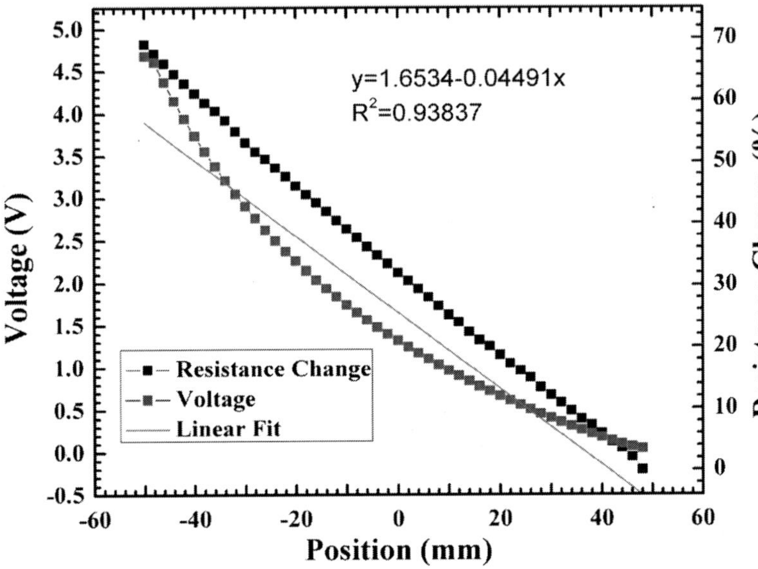

Figure 9. Coordinate plot and resistance change for a 0-to-5 V bias applied to the Ti metal line.

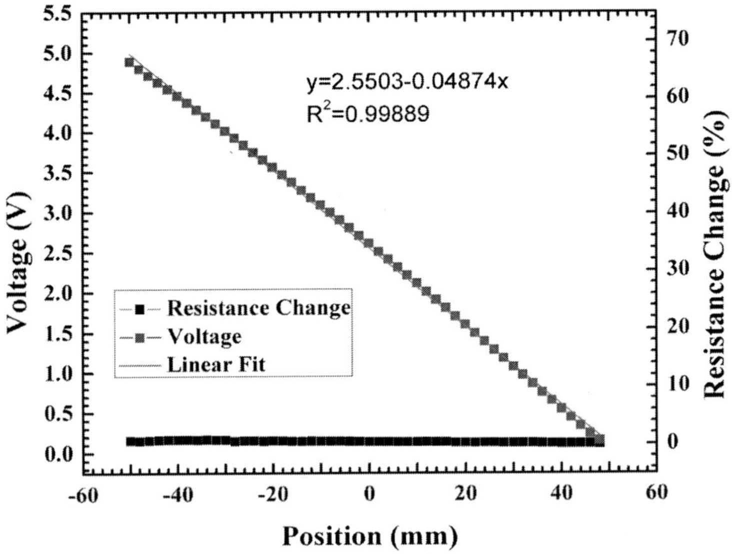

Figure 10. Coordinate plot and resistance change for a 0-to-5 V bias applied to the Ti/Au metal line.

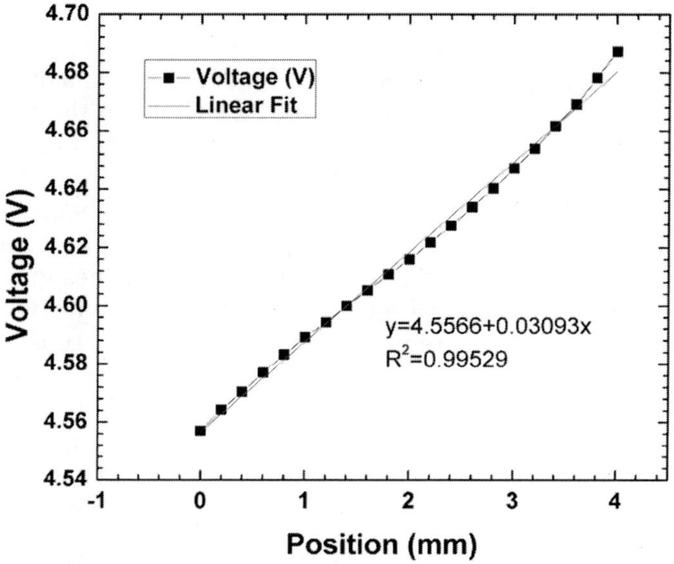

Figure 11. Coordinate plot over a 4-mm distance with 200 μm.

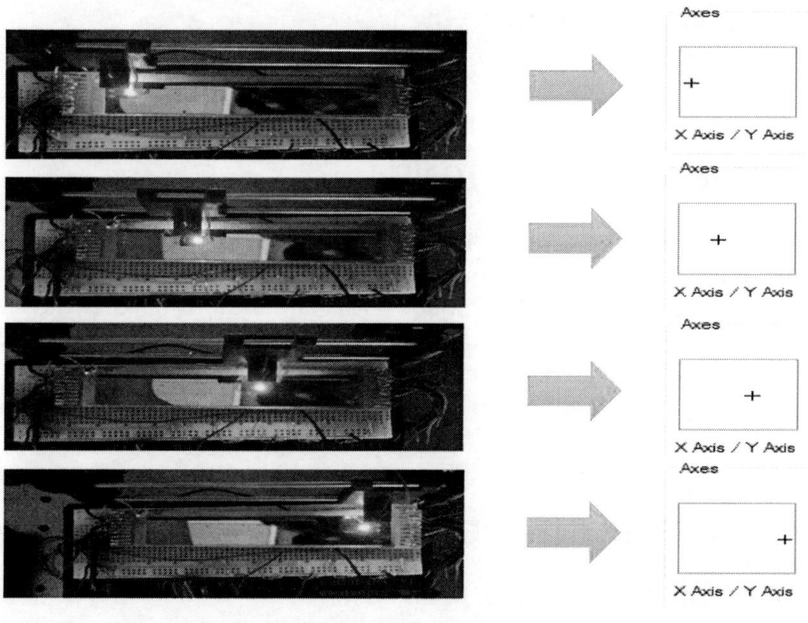

Figure 12. Demonstration of computer interfacing with SBPSD using built-in game controller configuration calibration settings.

4. CONCLUSIONS

This paper discusses a new approach to creating self-balancing position-sensitive detectors. The effect of resistance change on the linearity of the coordinate plot was investigated by comparing plots of voltage *vs.* light spot. Prototype devices were fabricated with lengths of 37 mm, 74 mm and 100 mm, respectively. Experimental measurements showed a linearity $R^2 = 0.9989$ and a spatial resolution of 200 μm for a 74-mm device in conditions where the photocurrent is 300 times greater than the dark current. A new type of position-sensitive device with linear coordinate characteristics requiring no external circuitry or backside contacts was fabricated and tested, and the performance characteristics of these devices were established, paving the way for ultimate integration into the target product as well as various other applications.

ACKNOWLEDGMENTS

The authors would like to acknowledge the New York State Trade Adjustment Assistance Center for partial funding and My Music Machines, Inc. (Schenectady, NY, USA).

AUTHOR CONTRIBUTIONS

Leigh Lydecker and Natalya Tokranova conceived and designed the experiments. Ryan Porrazzo performed, analyzed the data and wrote the paper. Suhasini Gattu helped with fabrication of the devices. Hasaram Bakhru helped with ion implanting the devices. James Castracane provided advisement for the project.

CONFLICTS OF INTEREST

The authors declare no conflict of interest.

REFERENCES AND NOTES

1. Wallmark, J.T. A New Semiconductor Photocell Using Lateral Photoeffect. *Proc. IRE.* **1957**, *45*, 474–483.

2. Zeng, L.; Yuan, F.; Song, D.; Zhang, R. A two-beam laser triangulation for measuring the position of a moving object.*Opt. Lasers Eng.* **1999**, *31*, 445–453.

3. Sharp Electronics. GP2D12/GP2D15-Distance Measuring Sensors. Product Data Sheet U970501. May 1997.

4. Makynen, A.; Kostamovaara, J. Digital Optical Position-Senstive Detector (PSD). In Proceedings of the 21st IEEE Instrumentation and Measurement Technology Conference (IMTC 04), Como, Italy, 18–20 May 2004; Volume 3, pp. 2358–2360.

5. Chowdhury, M.F.; Davies, P.A.; Lee, P. CMOS 1D and 2D N-Well Tetra-Lateral Position Sensitive Detectors. In Proceedings of the 1998 IEEE International Symposium on Circuits and Systems (ISCAS '98), Monterey, CA, USA, 31 May–3 Jun 1998; Volume 6, pp. VI-606–VI-609.

6. Mäkynen, A.; Ruotsalainen, T.; Kostamovaara, J. High accuracy CMOS position-sensitive photodetector (PSD).*Electron. Lett.* **1997**, *33*, 128–130.

7. Schottky, W. On the Origins of Photoelectrons in Cu_2 O-Cu Photocells. *Phys. Z Sowjetunion* **1930**, *31*, 913–925.

8. Lucovsky, G. Photoeffects in Nonuniformly Irradiated p-n Junctions. *J. Appl. Phys.* **1960**, *31*, 1088–1095.

9. Woltr, H.J. Single- and dual-axis lateral photodetectors of rectangular shape. *IEEE Trans. Electron. Devices* **1975**, *22*, 581–590.

10. Petersson, G.P.; Lindholm, L.-E. Position sensitive light detectors with high linearity. *IEEE J. Solid-State Circuits* **1978**,*13*, 392–399.

11. Martins, R.; Fortunato, E. Lateral photoeffect in large area one-dimensional thin-film position-sensitive detectors based in a-Si:H P-I-N devices. *Rev. Sci. Instrum.* **1995**, *66*, 2927–2934.

12. Martins, R.; Raniero, L.; Pereira, L.; Costa, D.; Águas, H.; Pereira, S.; Silva, L.; Gonçalves, A.; Ferreira, I.; Fortunato, E. Nanostructured silicon and its application to solar cells, position sensors and thin film transistors. *Philos. Mag.* **2009**,*89*, 2699–2721.

13. Andersson, H. Position Sensitive Detectors—Device Technology and Applications in Spectroscopy Applications in Spectroscopy. Ph.D Thesis, Mid Sweden University, Sundsvall, Sweden, 2008.

14. Kramer, J.; Baltes, H.; Seltz, P. Industrial CMOS technology for the integration metrology systems (photo-ASICs). *Sens. Actuators A Phys.* **1992**, *34*, 21–30.

15. Spieler, H. Introduction to Radiation Detectors and Electronics. Avaliable online: http://www-physics.lbl.gov/~spieler/ physics_198_notes/PDF/VI-PSD.pdf (accessed on 10 July 2015).

16. Radeka, V.; Rehak, P. Charge Dividing Mechanism on Resistive Electrode in Position-Sensitive Detectors. *IEEE Trans. Nuclear Sci.* **1979**, *26*, 225–238.

17. Hamamatsu Photonics, Chapter 02 Si Photodiodes. In *Hamamatsu Opto-Semiconductor Handbook*; Hamamatsu Photonics: Hamamatsu, Japan, 2014.

18. Bakker, M. De. The PSD Chip High Speed Acquisition of Range Images. Ph.D Thesis, Delft University of Technology, Delft, The Netherlands, 2000.

19. John, Wilson John, H. *Optoelectronics: An. Introduction*; Prentice Hall: London, England, 1998.

20. Hamamatsu Photonics. Si PIN Photodiodes S1223 Series Kpin1050E02 Datasheet. February 2013.

3

ANTI-REFLECTIVE NANO- AND MICRO-STRUCTURES ON 4H-SIC FOR PHOTODIODES

Min-Seok Kang, Sung-Jae Joo, Wook Bahng, Ji-Hoon Lee, Nam-Kyun Kim and Sang-Mo Koo

ABSTRACT

In this study, nano-scale honeycomb-shaped structures with anti-reflection properties were successfully formed on SiC. The surface of 4H-SiC wafer after a conventional photolithography process was etched by inductively coupled plasma. We demonstrate that the reflection characteristic of the fabricated photodiodes has significantly reduced by 55% compared with the reference devices. As a result, the optical response $I_{\text{illumination}}/I_{\text{dark}}$ of the 4H-SiC photodiodes were enhanced up to 178%, which can be ascribed primarily to the improved light trapping in the proposed nano-scale texturing.

1. INTRODUCTION

Up to now, silicon (Si) has been the dominant material for high-efficiency solar cells. However, Si-based devices perform well only under the limited conditions of relatively low temperatures and power ranges. Alternatively, in the research on wide-bandgap semiconductors, silicon carbide (SiC) has shown considerable potential for both high-power and optoelectronic devices [1]. SiC exhibits a wide-bandgap (3.26 eV) and superior thermal properties, which are advantageous for high-temperature applications and solar energy conversion [2]. However, polished SiC surfaces have a natural reflectivity with a strong spectral dependence. The reflectivity is inevitably high (20-40%), due to the high refractive index of n = 2.7-3.5 of SiC [3]. The optical losses associated with the reflectance of incident radiation are among the most important factors limiting the efficiency of a solar cell [4]. Therefore, photovoltaic cells normally require special surface structures or materials, which can reduce reflectance. A common solution is utilization of antireflection coatings based on interference, such as transparent layers of SiO_2 and Al_2O_3 [5]. However, such coatings worked only in a limited spectral range, and more efficient reflection reduction in a broad spectral range has been achieved by surface texturing, which can normally be accomplished by wet or dry etching. In principle, the wet etching of SiC can be done only with molten KOH at over 500°C, which is not a practical method. For that reason, dry etching with fluorine species, such as SF_6, and CF_4, is considered as the desirable method to form the textured surface of SiC [6].

In this article, we report a method for forming nano-scale-textured structures on 4H-SiC surfaces so as to reduce the surface reflectance of SiC. An inductively coupled plasma (ICP) etching was employed to form the structures, and the performance of the SiC photodiode cells was compared to that of reference cells without surface nano-scale texturing.

2. EXPERIMENTAL

Figure 1 shows the three different surface types of samples on 4H-SiC wafers that were prepared. In order to form nano-scale-textured honeycomb structures on the 4H-SiC surface, we first fabricated nano-structure patterns of the SiC surface. The samples were first cleaned in H_2SO_4:H_2O_2 = 4:1, followed by a BOE dip to remove the native oxide. The so-called nano-honeycomb etching process was performed in the following steps. First, to prepare a dry etching mask, a 100-nm Ni layer was sputtered and patterned by a conventional photolithographic process. A plasma-etching process was performed using SF_6 plasma (15% O_2 by flowing in a total gas load of 14 sccm) with ICP discharges at 550 W and RF chuck powers that created the dc self-bias from 117 V. The chamber pressure was 50 mTorr, and the sample was placed on the chuck that was cooled by He. Then, the remaining Ni was removed from the SiC surface by the Ni etchant (HF:H_2O_2:H_2O = 1:1:8). The honeycomb structures were created with a width and spacing, both of 3 μm, and a height of 100 nm as shown in Figure 2a. This method is used for forming the honeycomb structures of SiC surfaces which are referred to hereafter as *micro-honeycomb structures* [7, 8]. The substrate for SiO_2/4H-SiC was oxidized at 1150°C in O_2 for 5 h, and then a Si layer was deposited by electron-beam evaporation to be used as a masking layer for etching. The thicknesses of the SiO_2 and Si layers were 100 nm and 1 μm, respectively. Nano-scale texturing was performed using SF_6 plasma (17% O_2 by flowing in a total gas load of 24 sccm), with an ICP discharge power and a chamber pressure of 550 W and 30 mTorr, respectively, and a RF chuck power that created dc self-biases starting from 49 V. The nano-scale textures on the honeycomb structures were made by ICP etching as shown in Figure 2b, c [9]. This method is used for forming nano-scale-textured structures of SiC surfaces, referred to hereafter as *nano-honeycomb structures*, utilized the naturally roughened SiC surface morphology when the overlying Si turns into the so-called black Si by the ICP etching. After the black Si layer was consumed completely, the morphology was transferred to the underlying SiC, resulting in a roughened SiC surface.

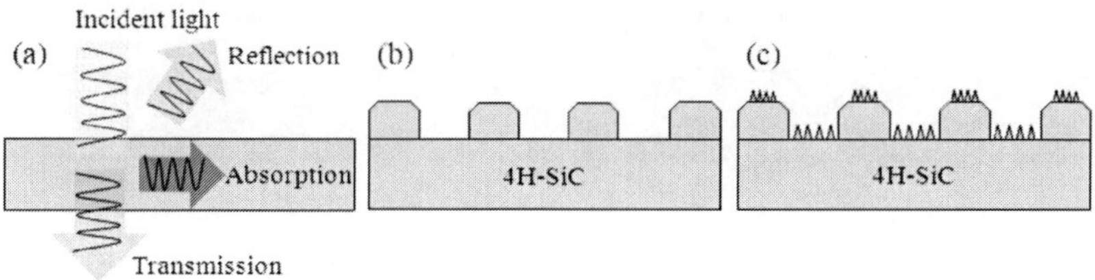

Figure 1 Schematic view of the 4H-SiC with different surface structures. **(a)** Reference cell, **(b)** micro-honeycomb structures, and **(c)** nano-honeycomb structures.

3. RESULTS AND DISCUSSION

Figure 2 shows scanning electron microscopy (SEM) images of the surface morphology of *nano-honeycomb structures*. Three different types of samples on SiC with different surface structures were examined: (a) reference structures, (b) micro-honeycomb structures, and (c) nano-honeycomb structures. The reflectance spectral dependence was studied using a UV-Vis/NIR spectrometer (AvaSpec-3648) and by AFM (N8 ARGOS) analysis. Figure 3 shows the corresponding reflectance spectra of the samples, along with those of the reference cells [10, 11]. In the region of wavelengths from 300 to 1000 nm, the reflectance of the*micro-honeycomb structures* was reduced by 30% with respect to that of the reference cell. After performing the unmasked ICP etching for additional nano-scale roughening on the *micro-honeycomb structures*, the reflectance decreased by 55% with respect to the reference cell. The optical measurements of the *nano-honeycomb structures* show that the amount of absorbed light significantly increased. The decreased reflectance of the structure is ascribed to the increased roughness of the surface due to the structures formed on the surface. Figure 4 shows the surface morphology observed with an atomic force microscope (AFM) under the contact mode with a scan area of 12 × 12 μm^2. The root mean square (RMS) of the surface roughness was calculated from the AFM images as shown in Figure 4d.

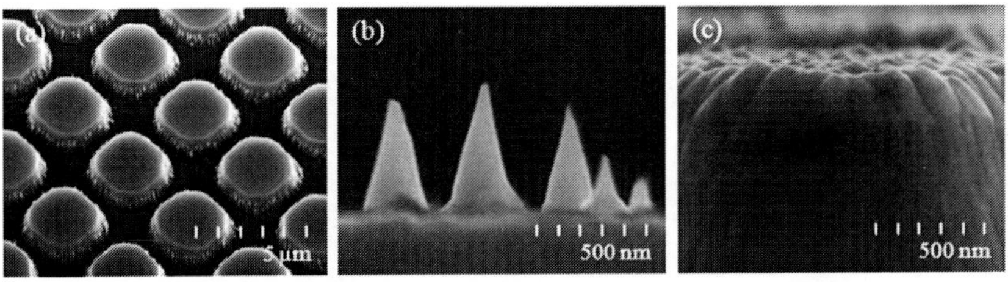

Figure 2 SEM images of representative "as-manufactured" structures. (a) The image shows the *nano-honeycomb structures* created by the photolithographic process. The detailed images show the rough surface on the bottom side **(b)** and the top side **(c)** of the *nano-honeycomb structures* created by the ICP-etching process using the gaseous mixture of $SF_6 + O_2$.

The relation between the reflectance and surface roughness can be described as [12]

$$\text{Reflectance of the surface} = r(1-p)/(1-pr) \tag{1}$$

where p represents the probability that, depending on the location on the rough surface, the incident photon is either absorbed with probability factor a, or reflected with a probability factor of $r = 1 - a$. As the surface roughness increases, the reflectance decreases, since more photons are absorbed. Similarly, as the RMS values of the *nano-*

honeycomb structures increases, the reflectance spectral dependence decreases because of the textured surface effect on the light trapping. It can be seen from the values of reflectance for 4H-SiC with different texturing structures that the *nano-honeycomb structures* exhibit clearly improved anti-reflective properties.

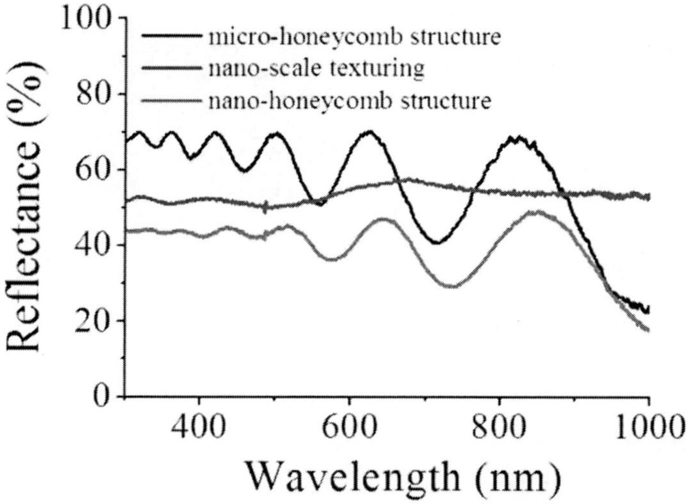

Figure 3 Comparison of spectral reflectivity from 300 to 1000 nm for different surface structures.

Schottky-type ultraviolet photodiodes were fabricated on *n*-type 4H-SiC wafers with a 12-μm-thick n^- epilayer ($N_D = 4.25 \times 10^{15}$cm^{-3}) grown on n^+ substrate ($N_D = 10^{18}$ cm^{-3}) [13]. A large area ohmic contact on the back-side was formed by the sputter of a 100-nm Ni film, followed by a rapid thermal annealing process at 950°C in N_2 for 90 s. The Schottky contacts on the front-side was fabricated by the electron-beam evaporation of a 50-nm Ni film, and a subsequent photolithographic patterning was performed to form rectangular ring patterns with widths of 550 μm and open area widths of 250 μm. Figure 5a shows the fabricated 4H-SiC Schottky photodiode structure. The open area directly exposed to radiation was estimated to be about 21% of the total device area. The current-voltage characteristics of the devices were measured by using a Keithley 4200 measuring unit. The saturated currents of the Schottky photodiodes were measured as a function of the reverse bias, both in the dark condition I_{dark} and under UV illumination at 300 nm $I_{illumination}$ [14]. Figure 5b compares the optical response ($I_{illumination}/I_{dark}$) of the photodiodes measured from the *micro-honeycomb structures* and *nano-honeycomb structures*.

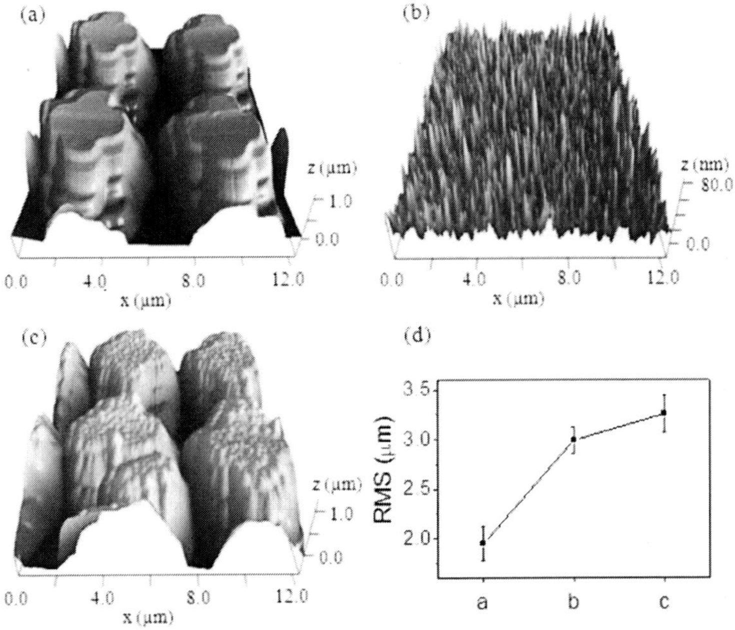

Figure 4 Contact-mode AFM images of 4H-SiC with different surface structures. **(a)** Micro-honeycomb structures, **(b)** nano-scale texturing, and **(c)** nano-honeycomb structures, as well as **(d)** RMS curve of the surface roughness.

The photocurrent shows a slight increase in the case of the *micro-honeycomb structures*, while a significant increase in optical response can be observed in the *nano-honeycomb structures* compared with the reference cell. The comparision of the photodiode properties for different structures are summarized in Table 1. For the reference cell, the measured I_{dark} and $I_{illumination}$ are 1.37×10^{-11} and 5.55×10^{-8} A, respectively, which results in the response of 75.4 A/W under the reverse bias of 20 V (see Table 1). The response values of 259.5 A/W at -

20 V were obtained at nano-honeycomb structures, as the optical reponse is increased by 178%. The optical response values at -20 V increased by 37 and 178% for *micro-honeycomb structures* and *nano-honeycomb structures*, respectively. The increased photocurrent gain is because the surface reflectance was reduced and the amount of absorbed light was increased with the *nano-honeycomb structures*. The results suggest that we can enhance the electro-optical response of the photodiodes by the anti-reflective effect of the nano-honeycomb-textured structures.

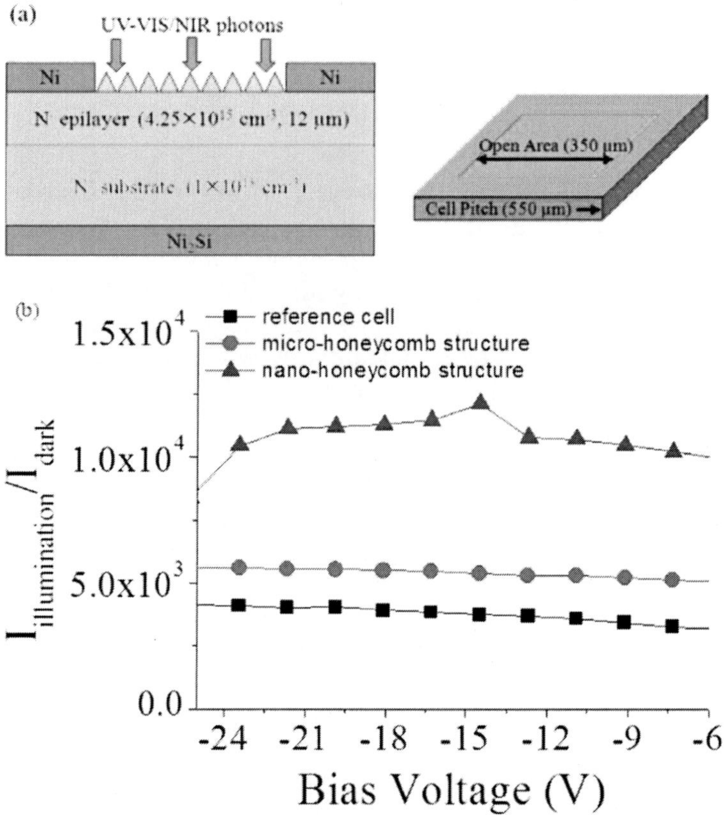

Figure 5 4H-SiC photodiode structure and the optical response characteristics. (a) Structure of the 4H-SiC Schottky-type photodiode with an open area of 250 × 250 μm^2. **(b)** Optical response of the 4H-SiC photo-diodes with different surface structures.

Table 1 Comparison of the Schottky-type ultraviolet photodiode properties for different structures

Structure	I_{dark} (A)	$I_{\text{illumination}}$ (A)	Response (A/W)
Reference cell	1.37×10^{-11}	5.55×10^{-8}	75.4
Micro-honeycomb	1.41×10^{-11}	6.32×10^{-8}	85.8
Nano-honeycomb	1.94×10^{-11}	2.18×10^{-7}	259.5

4. CONCLUSIONS

In summary, we proposed a method for fabricating nano-scale-textured structures on 4H-SiC surfaces to reduce reflection. After a conventional photolithography process to form the *nano-honeycomb structures*, the surface of 4H-SiC wafer was etched by ICP using a $SF_6 + O_2$ gas mixture. We demonstrated that the reflectance of the *nano-honeycomb structures* has significantly reduced by 55% compared with the reference cell. The reflectance was reduced because the roughness of the surface was increased. As a result, an optical response ($I_{\text{illumination}}/I_{\text{dark}}$) was increased by 178% for the *nano-honeycomb structures*, and an improved photocurrent was obtained from the subsequently fabricated 4H-SiC photo-diodes. The textured surface resulted in the reduction in reflectivity, which indicated that the amount of absorbed light increased because of efficient light trapping. It has been shown that the *nano-honeycomb structures* have proven as effective anti-reflective surface structures, which may open opportunities for the design of efficient photovoltaic cells on 4H-SiC.

ACKNOWLEDGEMENTS

This study was supported by the "System IC2010" project and "Survey of high efficiency power devices and inverter system for power grid" project of Korea Ministry of Knowledge Economy, by the National Research Foundation of Korea Grant funded by the Korean Government

2010-0011022, and by a Research Grant from Kwangwoon University in 2011.

COMPETING INTERESTS

The authors declare that they have no competing interests.

AUTHORS' CONTRIBUTIONS

MSK and carried most of the experiments. SJJ participated in the fabrication of micro- and nano-structures and analysis. WB and JHL performed the analysis of experimental data and measurement results. MSK prepared the manuscript initially. SMK conceived of the study and participated in its design and coordination. All authors read and approved the final manuscript.

REFERENCES

1. Liu X, Luo Z, Han S, Tang T, Zhang D, Zhou C: Band engineering of carbon nanotube field-effect transistors via selected area chemical gating. *Appl Phys Lett* 2005, 86: 243501. 10.1063/1.1944898

2. Guy OJ, Lodzinski M, Teng KS, Maffeis TGG, Tan M, Blackwood I, Dunstan PR, Al-Hartony O, Wilks SP, Wilby T, Rimmer N, Lewis D, Hopkins J: Investigation of the 4H-SiC surface. *Appl Surf Sci* 2008, 254: 8098. 10.1016/j.apsusc.2008.03.056

3. Koynov S, Brandt MS, Stutzmann M: Black nonreflecting silicon surfaces for solar cells. *Appl Phys Lett* 2006, 88: 203107. 10.1063/1.2204573

4. Manea E, Budianu E, Purika M, Cristea D, Cernica I, Muller R, Moagar V: Silicon solar cells technology using honeycomb textured front surface. *Sol Energy Mater Sol Cells* 2005, 87: 423. 10.1016/ j.solmat.2004.06.013

5. Zhang F, Yang W, Huang H, Chen X, Wu Z, Zhu H, Qi H, Yao J, Fan Z, Shao J: High-performance 4H-SiC based metal-semiconductor-metal ultraviolet photodetectors with Al_2O_3/**SiO_2** films. *Appl Phys Lett* 2008, 92: 251102. 10.1063/1.2949318

6. Leerungnawarat P, Hays DC, Cho H, Pearton SJ, Strong RM, Zetterling C-M, Östling M: Via-hole etching for SiC. *J Vac Sci Technol B* 1999, 17: 2050. 10.1116/1.590870

7. Zhao J, Wang A, Campbell P, Green MA: A 19.8% efficient honeycomb multicrystalline silicon solar cell with improved light trapping. *IEEE Trans Electron Dev* 1999, 46: 1978. 10.1109/16.791985

8. Lin CL, Chen PH, Chan CH, Lee CC, Chen CC, Chang JY, Liu CY: Light enhancement by the formation of an Al oxide honeycomb nanostructure on the n-GaN surface of thin-GaN light-emitting diodes. *Appl Phys Lett* 2007, 90: 242106. 10.1063/1.2748329

9. Joo SJ, Kang MS, Bahng W, Koo SM: Black SiC formation induced by Si overlayer deposition and subsequent plasma etching. *Thin Solid Films* 2011, 519: 3728. 10.1016/j.tsf.2011.01.279

10. Larruquert JI, Keski-Kuha RAM: Reflectance measurements and optical constants in the extreme ultraviolet for thin films of ion-beam-deposited SiC, Mo, Mg_2Si, and InSb and of evaporated Cr. *Appl Opt* 2000, 39: 2772. 10.1364/AO.39.002772

11. Haapalinna A, Kärhä P, Ikonen E: Spectral reflectance of silicon photodiodes. *Appl Opt* 1998, 37: 729. 10.1364/AO.37.000729

12. Berdahla P, Akbaria H, Jacobsb J, Klinkb F: Surface roughness effects on the solar reflectance of cool asphalt shingles. *Sol Energy Mater Sol Cells* 2008, 92: 482. 10.1016/j.solmat.2007.10.011

13. Jeong IS, Kim JH, Im SG: Ultraviolet-enhanced photodiode employing n-ZnO/p-Si structure. *Appl Phys Lett* 2003, 83: 2943. 10.1063/1.1615308

14. Sciuto A, Roccaforte F, Di Franco S, Raineri V, Billota S, Bonanno G: Photocurrent gain in 4H-SiC interdigit Schottky UV detectors with a thermally grown oxide layer. *Appl Phys Lett* 2007, 90: 223507. 10.1063/1.2745208

4

INFLUENCE OF NONPOLAR SUBSTANCES ON THE EXTRACTION EFFICIENCY OF SIX ALKALOIDS IN ZOAGUMHWAN INVESTIGATED BY ULTRA PERFORMANCE LIQUID CHROMATOGRAPHY AND PHOTODIODE ARRAY DETECTION

Yanling Zhao [1], Lei Jia [2], Hongbo Yang [3], Jiabo Wang [1], Ping Zhang [4], Ruisheng Li [5], Man Gong [2], Shengqiang Luo [2], Shijing Liu [2] and Xiaohe Xiao [1,*]

[1] China Military Institute of Chinese Medicine, 302 Hospital of People's Liberation Army, 100, the 4th Ring Road, Beijing 100039, China
[2] School of Chinese Medicine, The University of Hong Kong, Estates Building, 10 Sassoon Road, 999077, Hong Kong
[3] Yunnan Institute of Materia Medica, 22, Lengshuitang, Xishan District, Kunming, Yunnan 650111, China
[4] Department of Integrative Medical Center, 302 Hospital of the People's Liberation Army 100, 4th Ring Road, Beijing 100039, China
[5] Animal Laboratory Center, 302 Hospital of the People's Liberation Army, 100, 4th Ring Road, Beijing 100039, China

ABSTRACT

A reverse phase ultra performance liquid chromatography and photodiode array (UPLC-PDA) detection method was established for the determination of six alkaloids in Zoagumhwan (ZGW), and further for investigating the influence of nonpolar substances on the extraction efficiency of these alkaloids. The method was based on a BEH C_{18} (50 mm × 2.1 mm, 1.7 μm) column and mobile phase of aqueous phosphoric acid and acetonitrile including 0.05% buffer solution under gradient elution. ZGW samples of ZGW I, II, III and IV were obtained and prepared by pre-processing the crude materials of *Coptidis rhizoma* and *Evodiae fructus* using four technologies, namely direct water decoction, removal of nonpolar substances in *Evodiae fructus* by supercritical fluid extraction (SFE), removal of nonpolar substances in ZGW by SFE and removal of nonpolar substances in ZGW by steam distillation. The developed and validated UPLC-PDA method was precise, accurate and sensitive enough based on the facts that the six alkaloids showed good regression ($r > 0.9998$), the limit of detections and quantifications for six alkaloids were less than 28.8 and 94.5 ng/mL, respectively, and the recovery was in the range of 98.56%–103.24%. The sequence of the total contents of six alkaloids in these samples was ZGW II > ZGW IV > ZGW III > ZGW I. ZGW II, in which nonpolar substances, including essential oils, were firstly removed from *Evodiae fructus* by SFE, had the highest content of the total alkaloids, indicating that extraction efficiency of the total alkaloids could be remarkably increased after *Evodiae fructus* being extracted by SFE.

KEYWORDS

alkaloids; nonpolar substances; Zoagumhwan; UPLC-PDA; extraction efficiency

1. INTRODUCTION

Traditional Chinese medicines (TCMs) and their compound preparations have become more and more popular around the World during the last decade for their therapeutic effects which are complementary to Western medicines, and their capability to deal with many essential problems that have not yet been solved by current medicinal practices [1]. Zoagumhwan (ZGW) is a well-known compound preparation composed of *Coptidis rhizoma* and *Evodiae fructus* at the ratio of 6:1 (w/w), and has been used for more than 1,000 years. It is one of the simplest compound preparations but retains the basic therapeutic features [2].

Today, ZGW has been officially listed in the Chinese Pharmacopoeia [3] for its satisfactory effects of healing hypochondric and costal pain, stomach ache, acid reflux and nausea. It is reported [4,5,6,7] that the major active components of Zoagumhwan are alkaloids, namely columbamine, epiberberine, jatrorrhizine, coptisine, palmatine and berberine, as shown in Figure 1.

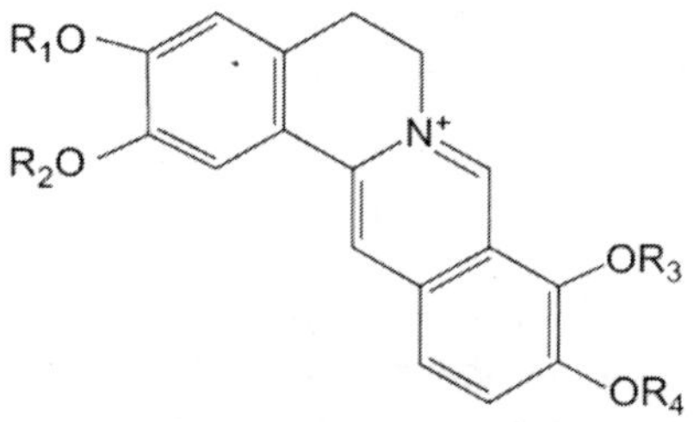

	R_1	R_2	R_3	R_4
Columbamine	CH_3	H	CH_3	CH_3
Epiberberine	CH_3	CH_3	$-CH_2-$	
Jatrorrhizine	H	CH_3	CH_3	CH_3
Coptisine	$-CH_2-$		$-CH_2-$	
Palmatine	CH_3	CH_3	CH_3	CH_3
Berberine	CH_3	CH_3	$-CH_2-$	

Figure 1. Structures of six tested alkaloids.

To date, many factors shown to affect the extraction efficiency of these alkaloids from Zoagumhwan, such as altering extraction solution [8], decoction conditions, sequence of decoction [9], *etc.*, have been investigated. However, it is well-known that *Evodiae fructus*, one of the components in Zoagumhwan, contains a volatile fraction, and monoterpenes and sesquiterpenes are the main volatile components in the volatile sample of *Evodia* according to gas chromatography-mass spectrometry (GC-MS) analysis [10,11,12]. Nevertheless, the influence of the nonpolar substances, including essential oils and other substances, on the extraction efficiency of the alkaloids was neglected in previous studies. In decoction, it is very easy that volatile components are freed from the herb. A small amount of them can be dissolved in the extraction solution, and this might affect the extraction of alkaloids from Zoagumhwan. Therefore, it is significant to investigate the influence of nonpolar substances on extraction efficiency of the main alkaloids.

Supercritical fluid extraction (SFE) and steam distillation are the most commonly utilized methods to extract nonpolar substances from medicinal and edible plant foods. In this study, the two methods were employed to remove the nonpolar fraction. Four different protocols, including direct water decoction, removal of nonpolar substances in *Evodiae fructus* by SFE, removal of nonpolar substances in Zoagumhwan by SFE and removal of nonpolar substances in Zoagumhwan by steam distillation were used to pre-process crude materials of *Coptidis rhizoma* and *Evodiae fructus* for obtaining different Zoagumhwan samples. Then, a rapid ultra performance liquid chromatography and photodiode array (UPLC-PDA) detection method was developed for the simultaneous determination of the six alkaloids in these extracts.

2. RESULTS AND DISCUSSION

2.1. Development of the UPLC Method

In the course of developing the rapid UPLC method, the basic chromatographic conditions such as mobile phase, flow rate, elution conditions, column temperature and detection wavelength were taken into consideration.

Mobile phase was a major factor for the good separation of the six tested alkaloids. Firstly, methanol, which was found to be less selective in the separation of the six alkaloids than acetonitrile, was discarded. According to the reports [13,14], the composition of acetonitrile and 20 mM ammonium acetate, which was adjusted to pH 10 by ammonia water, was investigated first. The results showed that the alkaloids could be separated well, but serious tailing was observed. Therefore, in order to improve the tailing, an acid solvent was considered according to the acid-base equilibrium theory. Choosing acetonitrile-0.5% (*v/v*) acetic acid could give an acceptable performance, but the separation between columbamine, jatrorrhizine, epiberberine and coptisine was not satisfactory. Finally, a more satisfactory chromatographic separation was acquired through the combination of acetonitrile and 0.05% (*v/v*) phosphoric acid water solution. A gradient elution procedure as shown in Table 1 was performed in order to decrease the total run time.

Attempts to further improve the chromatographic performance, the flow rate and the column temperature were adequately investigated. The experiments showed that the influence of column temperature was very remarkable on the separation of COL, JAT, EPI and COP. When the column temperature was raised from 15 °C to 40 °C, resolution was notablely degraded and the number of theoretical plates was raised notably. Finally, it was preferred to the column temperature of 22 °C.

Meanwhile, with a reduced flow rate, the tailing factor of the six alkaloids was increased.

Table 1. UPLC gradient elution procedure.

Time (min)	Flow rate (mL/min)	(A) 0.05% phosphoric acid water	(B) Acetonitrile
0.0	0.40	85%	15%
1.0	0.40	85%	15%
1.5	0.28	80%	20%
4.5	0.28	80%	20%
5.0	0.40	75%	25%
8.0	0.40	75%	25%

Three different UPLC-PDA detection wavelengths of 220, 270 and 340 nm were monitored and compared. 270 nm was finally selected for the lowest baseline noise for berberine, which was observed to be the best detection wavelength for all the subsequent experiments. Under the optimized UPLC conditions, the satisfactory and rapid separation of the six tested alkaloids shown in Figure 2 was achieved in 6 min with a back pressure of 7,800 psi.

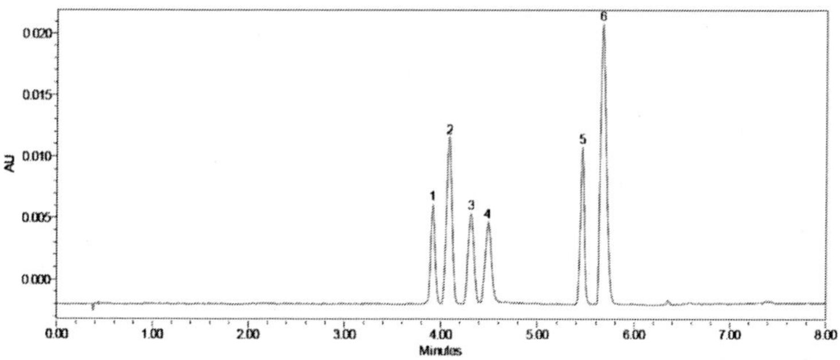

Peaks: 1. columbamine; 2. epiberberine; 3. jatrorrhizine; 4. coptisine; 5. palmatine; 6. berberine.

Figure 2. UPLC-PDA chromatograms of mixed standard solution.

2.2. Validation of the UPLC Method

The UPLC method was validated in terms of linearity, precision and accuracy according to ICH guidelines [15].

2.2.1. Calibration Curves

Methanol stock solutions containing six reference compounds were prepared and diluted to appropriate concentrations for the construction of calibration curves. At least six concentrations of the solution were analyzed in triplicate, and then the calibration curves were constructed by plotting the peak areas (y) *versus* the concentration (x) of each tested alkaloid. The results are shown in Table 2. The linearity of the calibration curves was verified by a correlation study and the correlation coefficients (r) were all better than 0.9998 within the test ranges.

2.2.2. Sensitivity

The methanol stock solution containing the six reference compounds was diluted to a series of appropriate concentrations with the same solvent, and an aliquot of the diluted solutions was injected into the UPLC system for PDA analysis. The limits of detection (LODs) and quantification (LOQs) under the present chromatographic conditions were determined at a signal-to-noise ratio (S/N) of about 3 and 10, respectively. Table 2 showed the data of LOD and LOQ for each investigated compound. The confidence LODs and LOQs for the six alkaloids were 25.8–31.8 and 85.0–104 ng/mL with 95% confidence level, respectively.

2.2.3. Precision

Intra- and inter-day variations were chosen to determine the precision of the developed method. For the intra-day variability test, six replicates of the mixed standards solution were analyzed within 1 day, while for the inter-day variability test, the solution was examined in duplicate for three consecutive days. For every calibration curve, the calibration concentrations were back-calculated from the peak area of the standard substances. The deviation from the nominal concentration was defined as accuracy. As shown in Table 3, variations, which were expressed by the intra- and inter-day relative standard deviations (RSD), were less than 0.9 and 1.2%, respectively.

Table 2. Linear regression data, LOD and LOQ of the six investigated alkaloids.

Analyte	Resolution [a]	Linear regression data			LOD	LOQ
		Regression equation [b]	Test range (μg/mL)	r	(ng/mL)	(ng/mL)
Columbamine	—	$y = 19154119x + 867.02$	0.19–31.20	0.9998	16.3	55.4
Epiberberine	2.02	$y = 19767219x - 628.2$	0.25–38.10	1.0000	25.8	92.9
Jateorrhizine	3.02	$y = 17377169x + 1454.2$	0.17–27.50	0.9998	21.5	72.0
Coptisine	1.53	$y = 18125088x - 91.1$	0.38–65.00	0.9999	22.3	76.3
Palmatine	10.61	$y = 10386051x - 699.2$	0.75–12.00	0.9999	19.6	64.3
Berberine	2.47	$y = 12063621x + 2611.3$	1.44–144.00	0.9999	28.8	94.5

[a] Resolution of the peaks of six alkaloids in UPLC profile was calculated using the Waters Empower 2 software; [b] y, peak area, x, concentration of each tested alkaloid.

2.2.4. Stability

With the temperature of both the laboratory and the control of the column and the auto-sample cell, being 22 °C, samples usually have to be stay at 22 °C for a long time, which should be considered carefully for unstable standard substances. In this study, the stability of the six alkaloids was tested by injecting standard solution at 0, 1, 2, 4, 8, 12 and 18 h at 22 °C. The results showed that all samples were stable under these conditions during the test time period (Table 3). Thus, the quantification of investigated compounds in all samples was preformed within 18 h.

2.2.5. Recovery

The recovery of the method was determined by spiking known amount of mixed standards in a certain amount (75 mg) of extract abtianed by direct water decoction. Three replicates were performed for the test. The mixture was extracted and analyzed using the method mentioned above. The results were calculated and are given in Table 4. The average recoveries of the five alkaloids were 98.56%–103.24%, with RSD values of less than 0.9%.

Table 3. Stability of the tested alkaloids, intra- and inter-day precision of the UPLC method.

Analyte	Stability (RSD, %)	Intra-day precision (n = 6)		Inter-day precision (n = 6)	
	22 °C	Accuracy (%) [a]	RSD (%)	Accuracy (%)	RSD (%)
Columbamine	1.0	100.1	0.8	100.4	1.2
Epiberberine	0.6	97.7	0.3	98.1	0.8
Jateorrhizine	0.8	101.3	0.9	100.8	0.9
Coptisine	0.8	98.0	0.7	98.5	1.0
Palmatine	0.8	99.6	0.6	99.7	1.2
Berberine	0.4	103.3	0.3	102.2	0.4

[a] Accuracy (%) = 100 × (mean measured concentration/nominal concentration).

Table 4. Recoveries for sample spiked with investigated alkaloids (n = 3).

Analyte	Original (μg)	Spiked (μg)	Found [a] (μg)	Recovery [b] (%)	RSD (%)
Columbamine	200.7	228.2	436.3	103.24	0.9
Epiberberine	288.9	324.5	619.5	101.88	0.8
Jateorrhizine	232.4	244.3	481.6	102.00	0.6
Coptisine	524.0	501.2	1018.0	98.56	0.8
Palmatine	816.8	885.2	1713.2	101.26	0.3
Berberine	2587.0	2474.6	5053.6	99.68	0.2

[a] Average of triplicates; [b] Recovery (%) = 100 × (amount found − original amount)/spiked amount.

All the above results illustrate that the developed UPLC-PDA method was precise, accurate and sensitive enough for simultaneous quantitative determination of the six alkaloids in Zoagumhwan samples.

2.3. Analysis of Real Samples

In this study, the contents of six alkaloids in four kinds of Zoagumhwan samples identified as ZGW I, II, III and IV from four different pre-processing methods were determined rapidly and simultaneously with the developed UPLC-PDA method. To reduce experimental error, each sample was extracted and analyzed in triplicate. The typical UPLC-PDA chromatograms of Zoagumhwan samples (ZGW I, II, III and IV) are shown in Figure 3. The target components were identified by comparing their retention times and UV spectra with those presented in the chromatogram of the mixture standard solution (Figure 2). The peak purity of target components from the four samples was verified using a photodiode array detector and found to be satisfactory. It could be seen from Figure 3 that the peak height and peak area of the same component had slight differences, though the number of predominant peaks from the four samples didn't change. Comparing the six characteristic peaks in the four chromatograms, one could find easily that peak 6 (berberine) is the most important active constituent in Zoagumhwan with the highest peak height.

In order to further find the differences between the same components in the four Zoagumhwan samples, the contents (mg/g) of the six investigated alkaloids in the four kinds of Zoagumhwan samples (ZGW I, II, III and IV) were determined in triplicate and listed in Table 5.

From the quantitative results, one could find that the contents of total alkaloids in the four samples showed significant differences. The sequence of the contents of total alkaloids in these samples was ZGW II > ZGW IV > ZGW III > ZGW I. ZGW II, in which nonpolar substances were firstly removed from *Evodiae fructus* by SFE-CO_2, had the highest content of total alkaloids. The results indicated that removal of nonpolar substances from *Evodiae fructus* could remarkably increase the extraction efficiency of the total alkaloids. In addition, the contents of total alkaloids in ZGW III and ZGW IV were lower than that in ZGW II. The reasons might be that parts of alkaloids in ZGW III were extracted by SFE-CO_2, and the extraction efficiency of alkaloids in ZGW IV was decreased by the remaining nonpolar substances. The increment, which was calculated by the formula: increment (%) = 100 × (the contents of total alkaloids in ZGW II, III or IV- ZGW I)/ZGW I, of ZGW II, III and IV were 10.9%, 0.8% and 4.3%, respectively.

On the other hand, the content of berberine was still the highest among the six major constituents in the samples, and the content of each alkaloid was influenced by nonpolar substances, but from Figure 4, it could be found that compared with the ZGW I sample, the content of columbamine and berberine was increased, while that of epiberberine and coptisine was decreased. The results indicated that the nonpolar substances existing in Zoagumhwan did not decrease the extraction yield of all alkaloids. The developed pre-processing method through removal of nonpolar substances in *Evodiae fructus* by SFE provided significant benefits in terms of extraction efficiency of the total alkaloids. Epiberberine, coptisine, palmatine and berberine all contain the $-OCH_2-$ group, which increases the dissolution rate under normal circumstance due to its polarity. However, with the absence of the role of the volatile oil when boiling, these four alkaloids in the solution vould reach new dissolution equilibrium. Epiberberin is the isomer of berberine, and would transfer into berberine for stability reasons. In the same way, the instability of $-OCH_2O-$ group would result in the decrease of content of coptisine. Thus, the above results were produced through a complicated interaction among chemicals, radicals, or ions in the extraction process. To confirm this deduction, further experiments should be conducted in the future.

3. EXPERIMENTAL

3.1. Materials and Reagents

Coptidis rhizoma and *Evodiae fruvtus* were purchased from Lv-Ye Pharm. Co. Ltd (Beijing, China) and authenticated by Professor Xiaohe Xiao (China Military Institute of Chinese Materia Medica, 302 Military Hospital of China). Columbamine hydrochloride (COL, No. 110711-200708), berberine hydrochloride (BER, No. 110717-200911), jateorrhizine hydrochloride (JAT, No. 110733-201108) and palmatine hydrochloride (PAL, No. 110732-201108) were supplied by the National Institute for the Control of Pharmaceutical and Biological Products in China. Coptisine hydrochloride (COP, No. 10/081025) and epiberberine sulphate (EPI, No. 10/081015) were supplied by Rong-He Pharm. Science and Technology Co. Ltd. (Shanghan, China). The purity of all reference compounds was more than 98.0%. Phosphoric acid was analytical purity from Beijing Chemical Works (Beijing, China). Methanol and acetonitrile were all of chromatographic purity from Fisher Chemicals (Pittsburgh, PA, USA). The water was double distilled water.

3.2. Instruments

A HA221-50-06 supercritical fluid extraction apparatus was purchased from Nantong Hua-An Supercritical Fluid Extraction Co. Ltd. (Nantong, Jiangsu, China). UPLC analysis was performed on a Waters Acquity Ultra performance liquid chromatography (UPLC) system (Waters Corp., Milford, MA, USA), equipped with a binary solvent delivery pump, an auto sampler and a photodiode array detector.

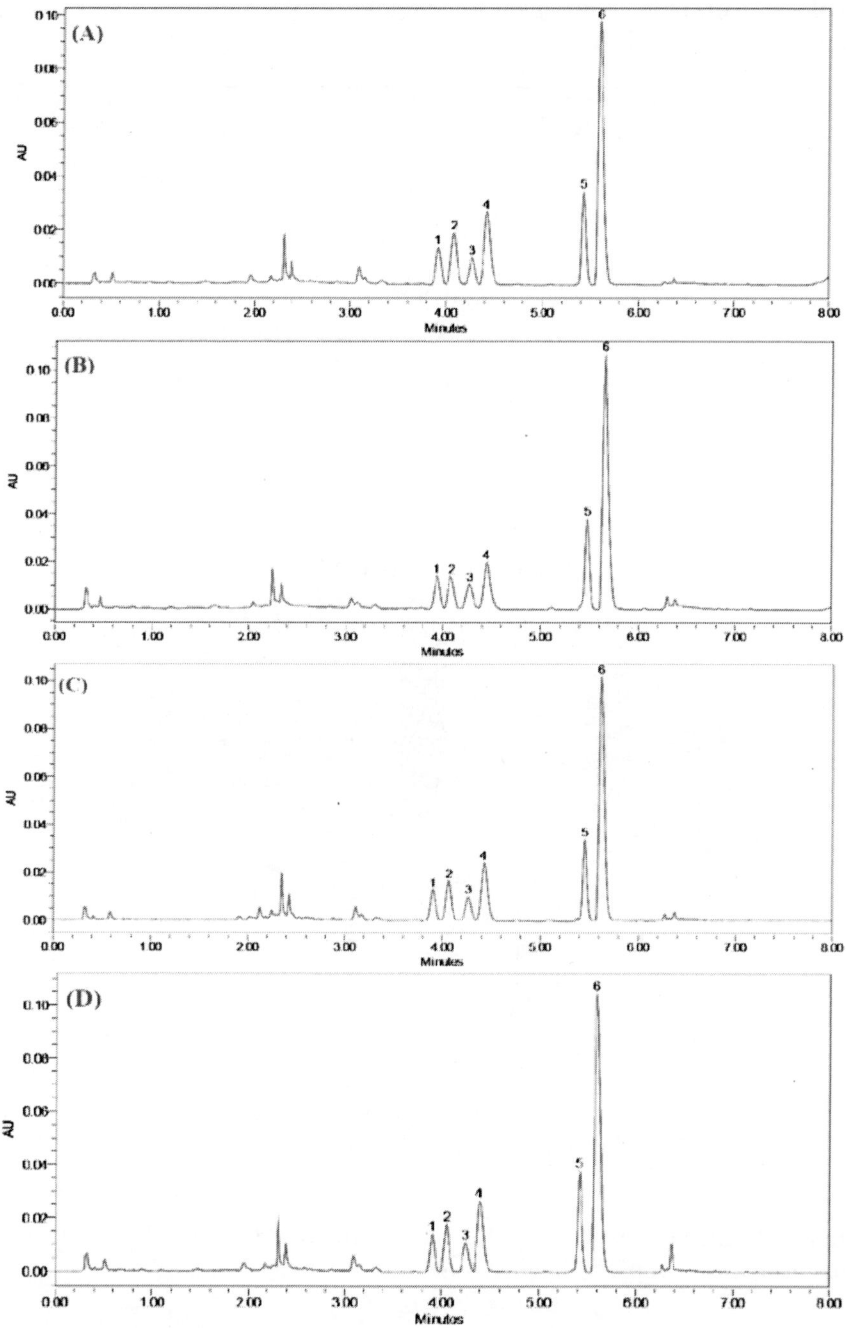

Figure 3. UPLC-PDA chromatograms of (**A**) ZGW I, (**B**) ZGW II, (**C**) ZGW III and (**D**) ZGW IV.

Table 5. Contents (mg/g) of investigated alkaloids in four kinds of Zoagumhwan samples.

Analytes	ZGW I [a]	ZGW II	ZGW III	ZGW IV
Columbamine	2.81 [b] ± 0.08	3.25 ± 0.08	3.01 ± 0.07	3.15 ± 0.09
Epiberberine	3.90 ± 0.07	2.96 ± 0.08	3.37 ± 0.06	3.27 ± 0.08
Jateorrhizine	2.70 ± 0.04	3.44 ± 0.05	2.72 ± 0.04	2.91 ± 0.04
Coptisine	6.85 ± 0.08	5.28 ± 0.10	5.86 ± 0.07	6.25 ± 0.09
Palmatine	11.02 ± 0.11	13.50 ± 0.13	10.69 ± 0.11	11.67 ± 0.09
Berberine	35.86 ± 0.72	41.58 ± 0.77	37.98 ± 0.68	38.58 ± 0.75
Total	63.14 ± 0.65	70.01 ± 0.71	63.63 ± 0.54	65.83 ± 0.69
Increment (%) [c]	——	10.9	0.8	4.3

[a] Zoagumhwan sample from four pre-processing methods on crude materials; [b] Average of triplicates; [c] Increment (%) = 100 × (the contents of total alkaloids in ZGW II, III or IV − ZGW I)/ZGW I.

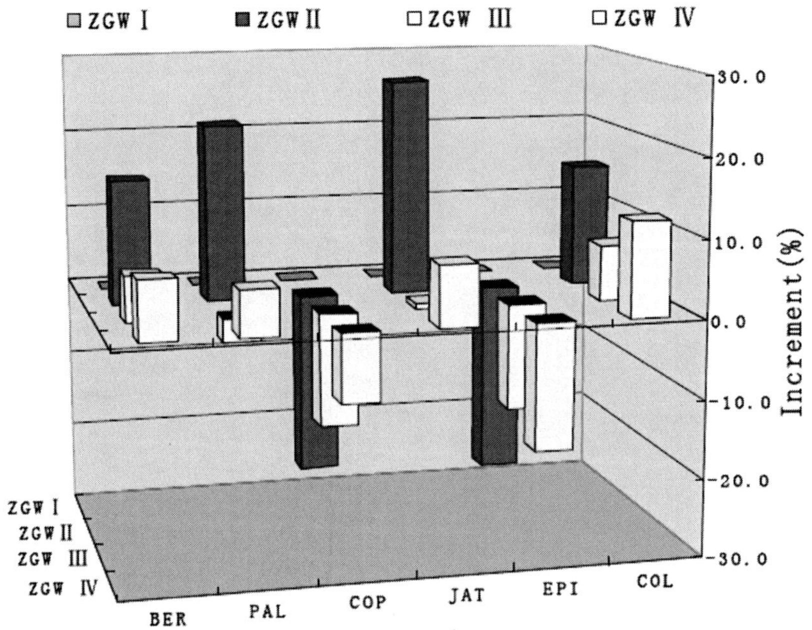

Figure 4. Influence of essential oil on the content of six alkaloids in Zoagumhwan samples.

3.3. Chromatographic Conditions

The chromatographic separation of six alkaloids was performed on a Waters Acquity BEH C_{18} (50 mm × 2.1 mm, 1.7 μm) column. The mobile phase consisted of phosphoric acid aqueous solution and acetonitrile, both including 0.05% (*v/v*) buffer solution. The gradient elution procedure has been listed in Table 1. The detection wavelength was set at 270 nm. The injection volume was 1 μL while the column was maintained at 22.0 ± 0.5 °C.

3.4. SFE Conditions

The SFE conditions of Zhang *et al.* [10] were used with slight modifications. The flow rate (CO_2) was kept at 4.7–15.7 L/h, extraction pressure was retained at 20–25 MPa, extraction time was 2 h, extraction temperature was maintained at 40.0 ± 2.0 °C and separation vessel II and vessel II were set 8 MPa and 6 MPa, respectively.

3.5. Preparation of Standard Solution

Mixed standard solution was prepared by dissolving six alkaloids, namely columbamine (7.8 mg), epiberberine (6.4 mg), jatrorrhizine (8.6 mg), coptisine (9.6 mg), palmatine (7.6 mg) and berberine (11.4 mg), with methanol (10 mL) in a flask. Then, the solution was transferred into a 50 mL volumetric flask, and made up to 50 mL. Finally, the standard stock solution was filtered through a 0.22 μm MILLEX GV syringe filter (Millipore, Bedford, MA, USA) prior to injection.

3.6. Pre-Processing of Crude Materials

3.6.1. Pre-Processing by Direct Water Decoction

Coptidis rhizoma and *Evodiae fructus* were crushed to pieces and mixed together thoroughly at a ratio of 6:1 (w/w). Then, the mixture was macerated in warm water (40 °C) for 30 min and decocted with water three times, 2 h for the first time, 1 h for the second time and 0.5 h for the third time, respectively. The filtrates from each decoction were blended and concentrated to a thick solution using a rotary evaporator (BÜCHI, Flawil, Switzerland) at 75 °C. The concentrated sample was dried in a vacuum oven at 50 °C, and the percentage yield was 26.25%.

3.6.2. Pre-Processing through Removal of Nonpolar Substances in Evodiae Fructus by SFE

Essential oil was removed firstly from *Evodiae fructus* by SFE under the above-described conditions. *Coptidis rhizoma* and*Evodiae fructus* without nonpolar substances at ratio of 6:1 (*w/w*) was mixed thoroughly and decocted with the same procedure as that of direct water decoction. The concentrated sample was dried in a vacuum oven at 50 °C, and the percentage yield was 25.72%.

3.6.3. Pre-Processing through Removal of Nonpolar Substances in Zoagumhwan by SFE

Coptidis rhizoma and *Evodiae fructus* were mixed thoroughly at a ratio of 6:1 (*w/w*). Then, essential oil in the mixture was removed by SFE under the above-described conditions. Zoagumhwan without nonpolar substances was decocted with the same procedure as that of direct water decoction. The concentrated sample was dried in a vacuum oven at 50 °C, and the percentage yield was 27.54%.

3.6.4. Pre-Processing through Removal of Nonpolar Substances in Zoagumhwan by Steam Distillation

Coptidis rhizoma and *Evodiae fructus* were thoroughly mixed first at a ratio of 6:1 (*w/w*). Then, the extraction method of steam distillation described in the Chinese Pharmacopoeia [16] was applied to remove nonpolar substances from the mixture. Zoagumhwan without essential oil was decocted with the same procedure as that of direct water decoction. The concentrated sample was dried in a vacuum oven at 50 °C, and the percentage yield was 24.58%.

3.7. Preparation of Zoagumhwan Sample Solutions

The concentrated and dried powders, namely ZGW I (78.75 mg), ZGW II (77.16 mg), ZGW III (82.62 mg) and ZGW IV (73.74 mg) from the above-described pre-processing methods of crude materials were dissolved in 10 mL methanol, shaken thoroughly and transferred into a flask. After being making up to 25 mL with methanol, the solutions were filtered through a 0.22 μm filter prior to UPLC analysis. For convenience, the sample solutions from the above-described four different pre-processing methods were labeled as ZGW I, II, III and IV, respectively.

3.8. Data Analysis

The Waters Empower software was used to collect and process the chromatograms data. Through importing the peak area and retention time in mean chromatograms, all parameters, including linear equation, correlations, and RSD, were calculated and analyzed using SPSS 13.0.

4. CONCLUSIONS

An easy and economical UPLC-PDA chromatographic procedure has been developed and successfully applied to determine the contents of six major alkaloids in Zoagumhwan extracts after different pre-processing technologies were used on the crude materials. The UPLC-PDA method was selective, sensitive and reproducible for the simultaneous determination of six alkaloids with low LOD and LOQ. The results have shown that the pro-processing technology through removal of nonpolar substances in *Evodiae fructus* by SFE could greatly improve the extraction efficiency of the main active components in Zoagumhwan. Thus, in the future, this described pro-processing method combined with UPLC analysis should be preferred and applied to pharmaceutical products of Zoagumhwan in order to improve the production technology, and to evaluate the productivity.

ACKNOWLEDGMENTS

The authors are grateful for the support from National Science Natural Foundation (No. 81173571), National Basic Research Program of China (973 Project) (2007CB512607) and Fund of The major projects of the National Science and Technology (2012ZX10005010-002-002).

REFERENCES AND NOTES

1. WHO. *General Guidelines for Methodologies on Research and Evaluation of Traditional Medicines*; WHO: Geneva, Switzerland, 2000; pp. 1–10.

2. Lin, C.L.; Hsu, H.Y. *Herb Couples*; Oriental Healing Arts Institute: Los Angeles, CA, USA, 1992; Volume 25, pp. 1–13.

3. National Commision of Chinese Pharmacopoeia. *Pharmacopoeia of the People's Republic of China*; Chemical Industry Press: Beijing, China, 2005; pp. 412–413.

4. Ko, Y.J.; Lee, J.S.; Park, B.C.; Shin, H.M. Inhibitory effects of Zoagumhwan water extract and berberine on angiotensinII-induced monocyte chemoattractant protein (MCP)-1 expression and monocyte adhesion to endothelial cells. *Vasc. Pharmacol.* **2007**, *47*, 189–196.

5. Wang, X.N.; Han, X.; Xu, L.N.; Yin, L.H.; Xu, Y.W.; Qi, Y.; Peng, J.Y. Enhancement of apoptosis of human hepatocellular carcinoma SMMC-7721 cells through synergy of berberine and evodiamine. *Phytomedicine* **2008**, *15*, 1062–1068.

6. Kong, W.J.; Zhao, Y.L.; Xiao, X.H.; Wang, J.B.; Li, H.B.; Li, Z.L.; Jin, C.; Liu, Y. Spectrum-effect relationships between ultra performance liquid chromatography fingerprints andanti-bacterial activities of *Rhizoma coptidis*. *Anal. Chim. Acta***2009**, *634*, 279–285.

7. Chu, C.Y.; Sheu, S.J. Liquid chromatographic separation of the alkaloids in coptis-evodia herb couple. *J. Chromatogr. A***1996**, *756*, 137–144.

8. Liu, S.T.; Zhao, X.H.; Huang, Y.Z. Effect of different solution to dissolution of five alkaloids in Zoagumhwan. *J. Nanjing TCM Univ.* **2007**, *23*, 172–174.

9. Pan, L.S.; Xu, X.M.; Lu, X.Y.; Wu, P.D. Study on contents variation of main chemical constituents during combination of *Rhizoma Coptidis* with *Fructus Evodiae* after separated decoction. *Chin. Pharmaceut. J.* **2005**, *140*, 258–261.

10. Zhang, Y.Z.; Lei, Z.J.; Wu, H.Q. Analysing supercritical fluid extraction of active component in *Evodia* by GC/MS. *J. Frist Milit. Med. Univ.* **2000**, *20*, S20–S21.

11. Lu, J.H.; Ning, D.S.; Tu, Y.S.; Jiang, B.; Zeng, Y.E. Assay of volatile oil from herb couple of *Rhizoma Coptidis* and *Fructus Evodiae* by GC/MS. *J. Guangzhou Univ. Tradit. Chin. Med.* **2003**, *20*, 312–315.

12. Pellati, F.; Benvenuti, S.; Yoshizaki, F.; Bertelli, D.; Rossi, M.C. Headspace solid-phase microextraction-gas chromatography-mass spectrometry analysis of the volatile compounds of *Evodia* species fruits. *J. Chromatogr. A* **2005**,*1087*, 265–273.

13. Kong, W.J.; Zhao, Y.L.; Xiao, X.H.; Jin, C.; Li, Z.L. Quantitative and chemical fingerprint analysis for quality control of*Rhizoma Coptidischinensis* based on UPLC-PAD combined with chemometrics methods. *Phytomedicine* **2009**, *16*, 950–959.

14. Chen, J.H.; Wang, F.M.; Liu, J. Analysis of alkaloids in *Coptis* chinensis Franch by accelerated solvent extraction combined with ultra performance liquid chromatographic analysis with photodiode array and tandem massspectrometry detections. *Anal. Chim. Acta* **2008**, *613*, 184–195.

15. ICH Topic Q2B. Validation of Analytical Procedures: Methodology (CPMP/ICH/281/95). Step4. Consensus Guideline. The European Agency for the Evaluation of Medicinal Products: Geneva, Switzerland, 1996.

16. National Commission of Chinese Pharmacopoeia, Appendix. In *Pharmacopoeia of the People's Republic of China*; Shao, M.L., Huang, J.F., Jiang, Z.J., Li, Z.J., Li, J.H., Yuan, Y.L., Ding, J.M., Yu, D.Q., Ma, D.L., Wang, P., *et al*, Eds.; Chemical Industry Press: Beijing, China, 2005; pp. 57–58.

5

INP-BASED WAVEGUIDE TRIPLE TRANSIT REGION PHOTODIODES FOR HYBRID INTEGRATION WITH PASSIVE OPTICAL SILICA WAVEGUIDES

Vitaly Rymanov *, Besher Khani, Sebastian Dülme, Peng Lu and Andreas Stöhr

Department of Optoelectronics/ZHO, University of Duisburg-Essen, Lotharstr. 55, Duisburg 47057, Germany

ABSTRACT

We report on a novel InP-based 1.55 μm waveguide triple transit region photodiode (TTR-PD) structure for hybrid integration with passive optical silica waveguides. Using the beam propagation method, numerical analyses reveal that, for evanescent optical coupling between a passive silica waveguide and the InP-based waveguide TTR-PD, a coupling efficiency of about 90% can be obtained. In addition to that, an absorption of about 50% is simulated within a TTR-PD length of 30 μm. For fabricated TTR-PD chips, a polarization dependent loss (PDL) of less than 0.9 dB is achieved within the complete optical C-band. At the operational wavelength of 1.55 μm, a reasonable PDL of 0.73 dB is measured. The DC responsivity and the RF responsivity are achieved on the order of 0.52 A/W and 0.24 A/W, respectively. Further, a 3 dB bandwidth of 132 GHz is experimentally demonstrated and high output-power levels exceeding 0 dBm are obtained. Even at the 3 dB cut-off frequency, no saturation effects occur at a photocurrent of 15.5 mA and an RF output power of −4.6 dBm is achieved. In addition to the numerical and experimental achievements, we propose a scheme for a hybrid-integrated InP/silicon-based photonic millimeter wave transmitter.

KEYWORDS

photonic integrated circuits; silicon photonics hybrid integration; heterogeneous integration; indium phosphide photonics; silica and silicon nitride photonics; optical interconnects; telecommunications; microwave photonics

1. INTRODUCTION

Research activities targeting millimeter wave (mmW) applications, *i.e.*, applications utilizing electromagnetic waves within the frequency range from 30 GHz to 300 GHz, have been revived in recent years. Many modern applications in sensing [1], imaging [2], radar [3], and, most notably, in wireless communications [4,5], make use of ever higher frequencies. One prominent example is that of future 5G mobile systems, which are currently being considered to make use of carrier frequencies in the mmW range [6,7]. This has simultaneously revived the interest in solid-state components and it has led to scores of high-performance components enabling, for example, highly-efficient mmW signal generation and detection [8]. In addition to purely electrical solutions using advanced InP- or SiGe-based solid-state amplifiers and mixers, integrated photonic technology has also become more mature, leading to photonic solutions ready for real-life implementations, e.g., the generation and distribution of low-phase-noise mmW local oscillator signals in radio astronomy [9]. Additionally, research on broadband wireless communications has led to recent outstanding results using photonic technologies. In particular, radio-over-fiber (RoF) techniques have enabled the transmission of extreme high-data (100 Gbit/s) wireless signals within mmW bands [10].

High-speed photodiodes (PDs) are key in such broadband RoF wireless networks. To meet the challenging demands for advanced mmW PDs in terms of bandwidth or output power, various groups are researching evolutionary, but also revolutionary, concepts helping to improve the performances of state-of-the-art mmW PDs [1,5,8,11,12,13,14,15,16,

17,18]. Key objectives of this research are to expand the operational bandwidth of the PDs and simultaneously attain higher power-saturation levels [1]. High-frequency operation typically leads to PD structure designs including relatively thin absorptive layers to enhance the transit-time-limited response, which on the other hand simultaneously deteriorates the external quantum efficiency, primarily in PDs with top-illumination [13,14]. Furthermore, the resistance-capacitance (RC) time constant becomes the dominating limitation in terms of bandwidth for thin, undoped absorbing layers. The necessity to simultaneously achieve high external quantum efficiencies, high operational bandwidths for broadband (or full-band) operation, as well as high radio frequency (RF) output-power levels, favor the use of waveguide-based PDs [15,16]. This is due to their flexibility in the layer structure and topology design. As an example, mushroom-type waveguide PDs offer a solution for improving RC limitations, enabling high-frequency operation [17].

In this work, we report on InP-based 1.55 μm triple transit region PDs (TTR-PDs) [18], which employ the waveguide approach for hybrid integration with passive optical silica waveguides. As a future application (e.g., for 5G mobile systems), we first propose a scheme of a hybrid-integrated photonic mmW transmitter for photonic RF beam forming and beam steering. Since this approach is based on heterogeneous integration of InP- and silicon-based photonics components, the optical waveguide coupling is to be investigated for connecting the optical beam-steering network (including passive optical silica waveguides) [19,20] with a linear TTR-PD array. One of the challenges of these hybrid-integrated waveguide devices remains in the design of an appropriate layer structure, which enables efficient coupling of the optical mode from the passive silica waveguide into the active InP-based waveguide. To study evanescent coupling, we introduce an active waveguide based on the TTR-PD layer structure [18] on top of a passive optical silica waveguide [21]. Numerical simulations using the beam propagation method (BPM) are accomplished to demonstrate the optical TE-mode propagation in the hybrid-integrated InP/silica double waveguide. For experimental characterization, several TTR-PD chips were fabricated with monolithically-integrated passive optical waveguides to couple light into the active PD structure and, thus, measure the responsivity (DC and RF), the 3 dB bandwidth, and the 1 dB compression point of the TTR-PDs prior to further hybrid integration with silica waveguides. Accordingly, we report high-power (>0 dBm) TTR-PDs with a measured 3 dB bandwidth beyond 130 GHz.

2. SCHEME OF A HYBRID-INTEGRATED INP/SILICON-BASED PHOTONIC MILLIMETER WAVE TRANSMITTER

For future 5G mobile systems considered to make use of carrier frequencies in the mmW range, e.g., in the 60 GHz or the 70/80 GHz frequency band [4,6,7], photonic mmW transmitters that simultaneously enable high output-power levels, high operational bandwidth, and the opportunity for photonic beam forming, as well as beam steering, are necessitated. For this reason, a hybrid integration of different solid-state technologies on one platform becomes key for such high-performance devices. One of the benefits for a hybrid-integrated waveguide PD would be, for example, the low propagation loss (<0.1 dB/cm) [21] of the passive optical silica waveguides used in optical beam-steering networks [19,20]. A schematic view of a hybrid-integrated InP/silicon-based photonic mmW transmitter is illustrated in Figure 1.

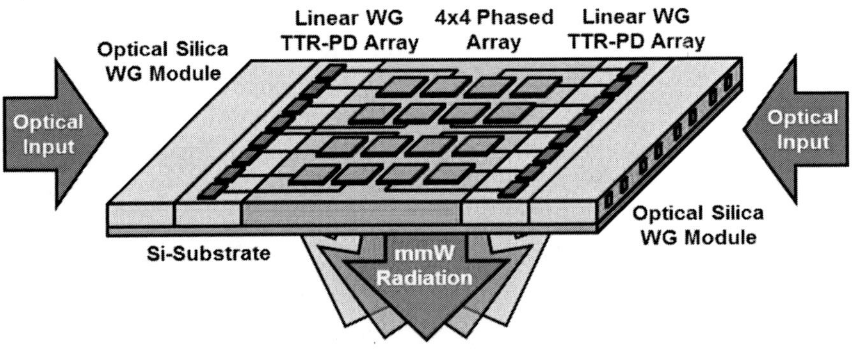

Figure 1. Schematic view and operation principle of a hybrid-integrated InP/silicon-based photonic millimeter wave (mmW) transmitter featuring two optical silica waveguide (WG) modules, two linear WG TTR-PD arrays and a 4 × 4 phased array antenna for mmW beam forming and beam steering.

As depicted in the scheme, the proposed photonic mmW transmitter is composed of different optics (optical waveguide modules), optoelectronics (waveguide TTR-PD arrays), and RF (phased array) sections. Two optical silica waveguide modules are integrated for double-sided illumination at the optical wavelength of 1.55 μm. An optical fiber array with eight conventional (*i.e.*, cleaved) single-mode fibers (SMFs) is considered for coupling of light into each silica module, which comprises monolithically-integrated spot-size converters [21] and an optical beam-steering network, e.g., using the true-time-delay approach [19,20].

The optical inputs of the linear TTR-PD array are interfaced with the corresponding optical outputs of the passive silica waveguides from the silica module. To couple the light vertically into the active TTR-PD structure, the TTR-PD is to be integrated on top of the passive silica waveguide. In the TTR-PD, the optical signal with the appropriate beam-steering signal (*i.e.*, phase) gets absorbed and converted into an electrical (mmW) signal that is further fed to an integrated antenna element, which is located on a silicon platform. Each TTR-PD allows individually addressing an antenna element of the phased array. Similar integration approaches on silicon submounts were already reported in previous works on hybrid-integrated InP/silicon-based photonic transmitter modules employing highly-resistive silicon spacers and hemispherical lenses [22,23]. Finally, the relative phases of the generated mmW signals can be adjusted in such a way that an efficient mmW beam of the entire phased array is generated and steered in a preferential direction [19,20].

3. SIMULATION OF EVANESCENT OPTICAL COUPLING BETWEEN THE TTR-PD AND A PASSIVE SILICA WAVEGUIDE

As the TTR-PD is a key element in the photonic transmitter scheme, which converts the optical signal from the optical to the RF domain, evanescent optical coupling between a silica (SiO_2) waveguide and the TTR-PD was studied using beam propagation method (BPM) analyses.

3.1. Modeling of the Hybrid-Integrated InP/Silica Double Waveguide

The cross section of the hybrid-integrated InP/silica double waveguide, which consists of an active InP-based TTR-PD and a passive silica waveguide section, is depicted schematically in Figure 2. In contrast to the monolithic approach that was reported in [18], the mushroom-type TTR-PD structure is flipped and hybrid-integrated on top of the silica waveguide. In addition to that, the thickness of the highly p-doped InP layer is reduced from 800 nm (see [18]) to 400 nm. Here, it should be noted that the free carrier absorption in the highly p-doped InP cladding layer gets higher than in the slightly n-doped InP cladding layer of the monolithically-integrated InGaAsP/InP waveguide (see Section 3.2 and Section 4).

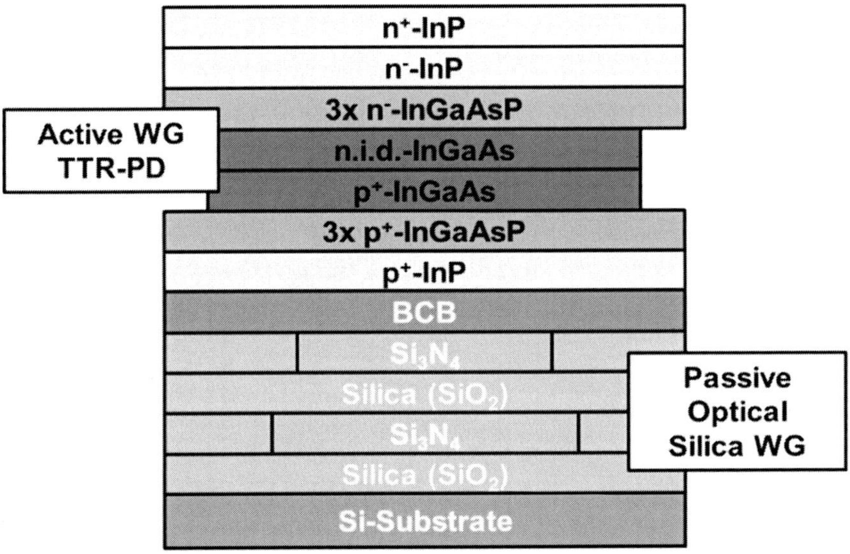

Figure 2. Schematic cross section of the developed InP/silica double waveguide comprising an active InP-based waveguide (WG) TTR-PD with a hybrid-integrated passive optical silica WG. Benzocyclobutene (BCB) is used for adhesive wafer bonding.

Further, the active TTR-PD structure comprises a 110 nm thick InGaAs core featuring a non-intentionally doped (n.i.d.) absorber layer (70 nm) and a highly p-doped absorber layer (40 nm) that are incorporated into the InGaAsP spacers and the InP cladding layers. The passive silica waveguide features two embedded silicon nitride (Si_3N_4) stripes and is grown on a Si-substrate. Details on both layer structures (*i.e.*, TTR-PD and silica waveguide) have been reported in [18,21]. To connect both waveguides, adhesive wafer bonding can be applied introducing a thin (<0.1 μm) intermediate benzocyclobutene (BCB) layer as BCB provides a refractive index of ~1.5 that is similar to the refractive index of silica (~1.45).

3.2. Numerical Beam Propagation Method Simulations

Based on the proposed layer structure from Figure 2, numerical BPM simulations for the propagation of TE-polarized light were accomplished for the developed InP/silica double waveguide (see Figure 3). To calculate the effective length of the TTR-PD with respect to coupling efficiency and absorption, the active coupling length was set to 100 μm.

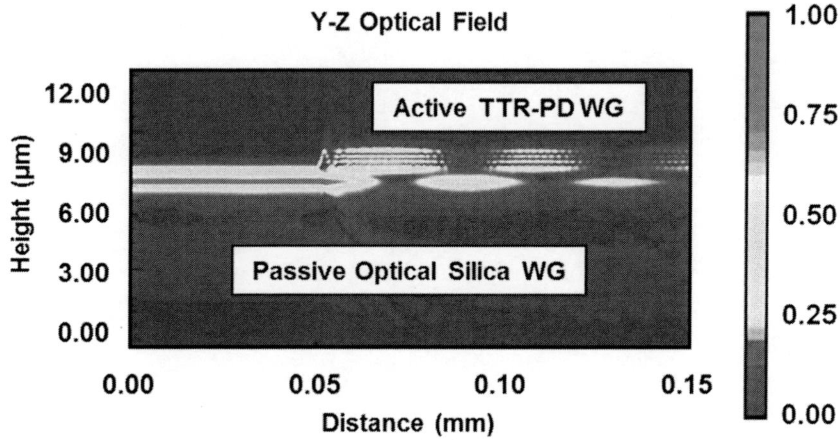

Figure 3. Simulated propagation of TE-polarized light demonstrating evanescent coupling between a silica waveguide (WG) and the InP-based TTR-PD WG (longitudinal cut).

From Figure 3, one can observe that a coupling efficiency of more than 90% within a coupling length of about 25 μm is achieved. It can further be observed that an absorption of about 50% is obtained in the InGaAs layers before the light couples into the silica waveguide (see Figure 3 and Figure 4). Additionally, the BPM simulations have indicated that around 14% of the propagation loss can be traced back to the free carrier absorption in the highly p-doped InP layer, which somewhat deteriorates the external quantum efficiency but actually does not affect the 3 dB bandwidth of the TTR-PD as the n.i.d.-InGaAs absorber layer, the slightly n-doped InGaAsP spacers, and the slightly n-doped InP collector of the TTR-PD remain unmodified, such as presented in [18].

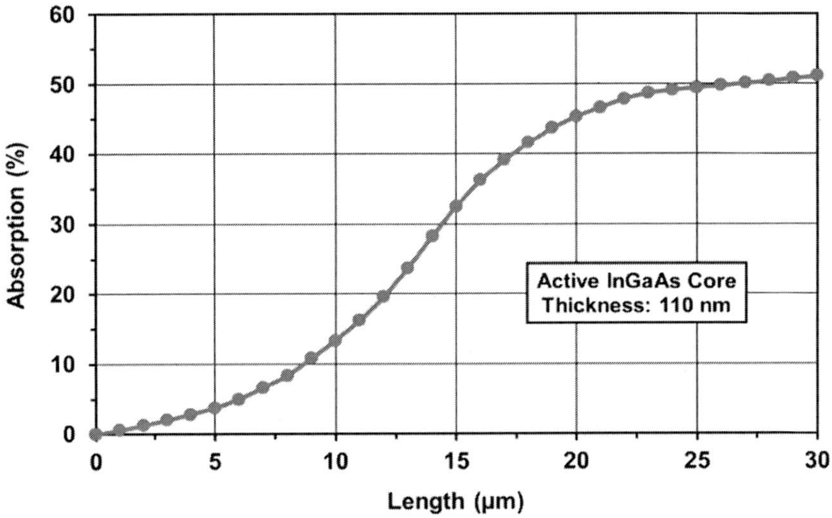

Figure 4. Simulated absorption at the optical wavelength of 1.55 μm plotted *versus* the length of the TTR-PD waveguide with an InGaAs core width of 2 μm.

It is worth mentioning that an absorption of around 80% could be achieved within the simulated coupling length of 100 μm, even though the absorption does not take place over the entire propagation term (see Figure 3). However, the extended length would raise the TTR-PD's junction capacitance, which is directly proportional to the length of the TTR-PD and, thus, deteriorates the 3 dB bandwidth.

4. EXPERIMENTAL RESULTS OF FABRICATED WAVEGUIDE TRIPLE TRANSIT REGION PHOTODIODE CHIPS

Prior to interfacing the TTR-PDs with integrated silica waveguides, the DC responsivity, the RF responsivity, the 3 dB bandwidth, and the 1 dB compression point of the developed TTR-PD structure were to be investigated. Based on the achieved BPM simulation results, several devices with an active TTR-PD area of 2 × 30 μm^2 were fabricated and experimentally characterized. To characterize the TTR-PDs, a monolithically-integrated passive optical InGaAsP/InP-based waveguide, providing a similar optical TE-mode profile with a mode field diameter (MFD) on the order of 1.5 μm [21], was substituted for the silica waveguide in the layer structure design to enable (evanescent) optical coupling to the TTR-PD chip (see Figure 5a). The top view of a fabricated TTR-PD chip is further shown in Figure 5b. For on-chip measurements, no additional cooling arrangements, such as Peltier elements or heat sink modules, were applied to the TTR-PD chips. Further, it should be noted that no anti-reflection coating was deposited at the passive optical waveguide facets.

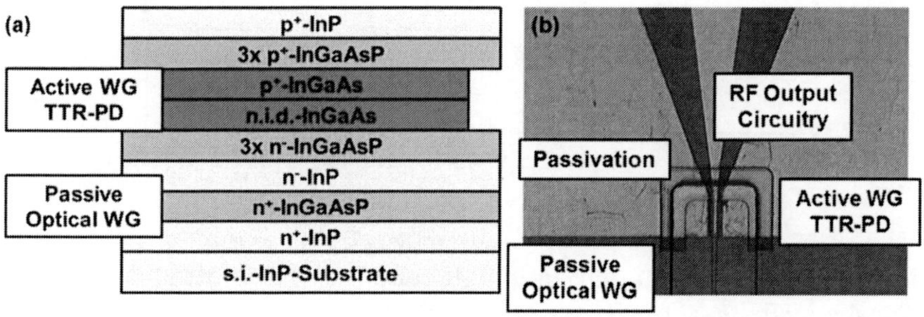

Figure 5. (**a**) Schematic cross section of the TTR-PD layer structure with a monolithically-integrated passive optical waveguide (WG) and (**b**) scanning electron microscope (SEM) photograph of the fabricated WG TTR-PD chip.

4.1. Polarization Behavior and DC Responsivity

To investigate the polarization behavior for optical heterodyne mmW signal generation (see Section 4.2), the polarization dependent loss (PDL) of the TTR-PDs was measured within the complete optical C-band from 1530 to 1565 nm (see Figure 6). The optical fiber-chip coupling to the passive optical waveguide was accomplished using a lensed SMF. The applied SMF featured a spot size diameter on the order of 1.7 μm to reduce the optical mode mismatch, which would be induced by using a standard SMF with an MFD of about 10.4 μm. For measuring the PDL, the optical input-power level was set to 10 dBm. Two direct current (DC) needle manipulators were used for biasing the TTR-PD chip at a bias voltage of −8 V. The measured PDL is plotted in Figure 6.

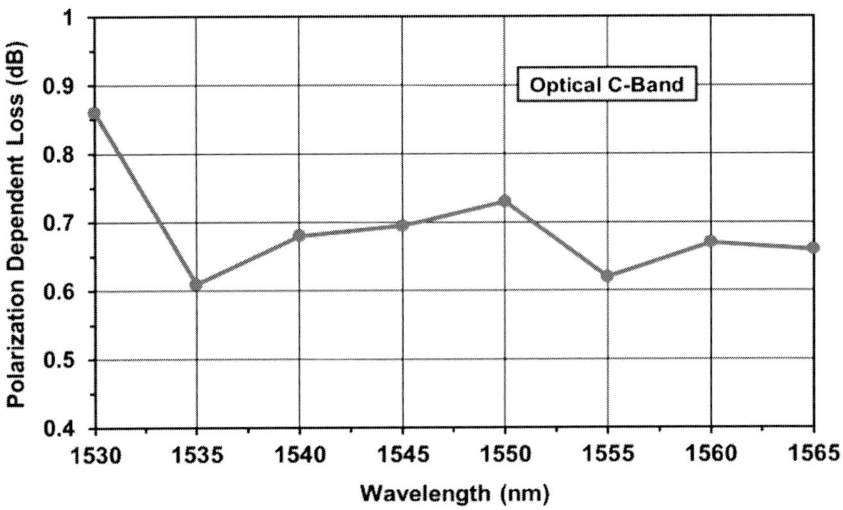

Figure 6. Measured polarization dependent loss plotted against the optical wavelength for a bias voltage of −8 V.

As can be seen from Figure 6, within the entire C-band the TTR-PDs exhibit a reasonable PDL, which is less than 0.9 dB. A PDL of 0.73 dB can further be observed at the operational wavelength of 1.55 μm. In addition to that, a peak DC responsivity ($R_{DC,max}$) on the order of 0.52 A/W was achieved (see Figure 7).

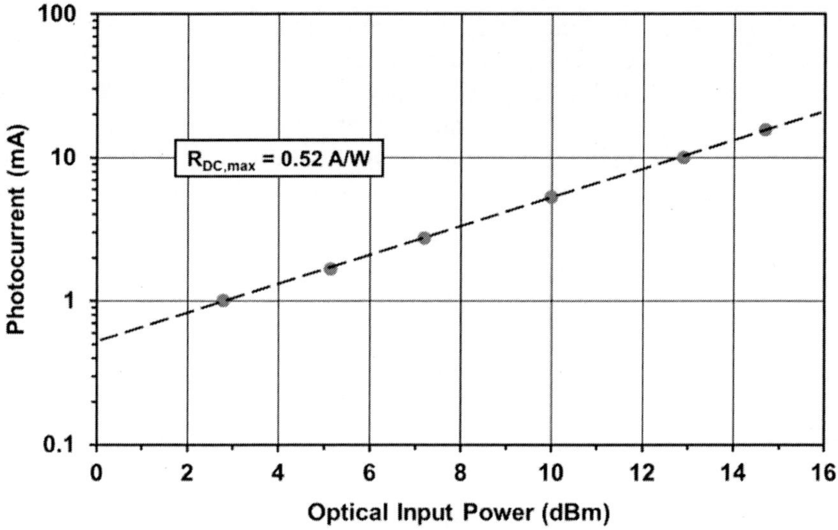

Figure 7. Measured photocurrent *versus* the optical input power for the operational wavelength of 1.55 μm at a bias voltage of −8 V. The dashed line shows the simulated photocurrent for the DC responsivity of 0.52 A/W.

4.2. Measured Frequency Response

For measuring a frequency response from DC up to 220 GHz, we used a calibrated scalar network analyzer setup featuring a tunable external cavity laser (TLS) in conjunction with a polarization controller (PC) and a continuous wave laser source (LS) for optical heterodyne mmW signal generation. The measurement setup is illustrated in Figure 8. Three different 100 μm pitch coplanar RF probes (~50 Ω) were used to attain on-chip power measurements within the measured frequency range shown in Figure 9.

Up to a frequency of 110 GHz, an RF probe and an external bias-T, both including a 1 mm coaxial W1-connector, were applied together with an electrical spectrum analyzer (ESA) employing an internal mixer (up to 50 GHz) and the external V-band (50–75 GHz) and W-band (75–110 GHz) harmonic mixers. Within the frequency ranges of 90–140 GHz (F-band) and 140–220 GHz (G-band), two further RF probes featuring a WR-08 and a WR-05 waveguide output, respectively, and an integrated bias-T were used with corresponding Schottky barrier diodes (SBDs) and a volt meter. The measured frequency response up to 190 GHz is presented in Figure 9. The photocurrent and the bias voltage of the TTR-PD were set to 1 mA and −8 V, respectively. From Figure 9, one can notice that a 3 dB cut-off frequency of 132 GHz is achieved. The resulting RF responsivity is found to be on the order of 0.14 A/W. At lower frequencies, a maximum RF responsivity of about 0.24 A/W has been achieved.

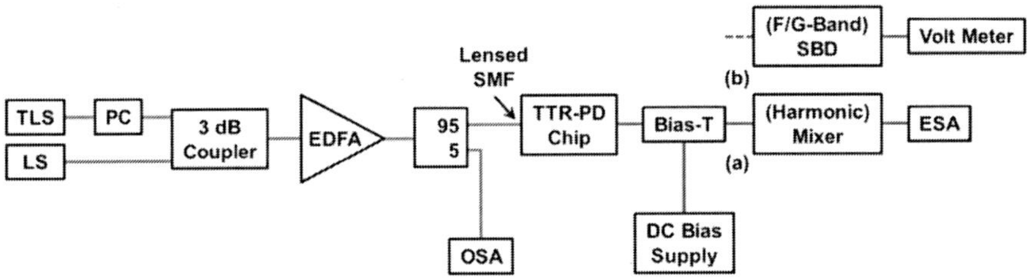

Figure 8. Measurement setup based upon optical heterodyne millimeter wave signal generation up to 220 GHz featuring (**a**) an electrical spectrum analyzer (ESA) with internal and external mixers (DC–110 GHz) and (**b**) Schottky barrier diodes (SBDs) connected to a volt meter (90–220 GHz). An erbium-doped fiber amplifier (EDFA) and an optical spectrum analyzer (OSA) are further included in the setup.

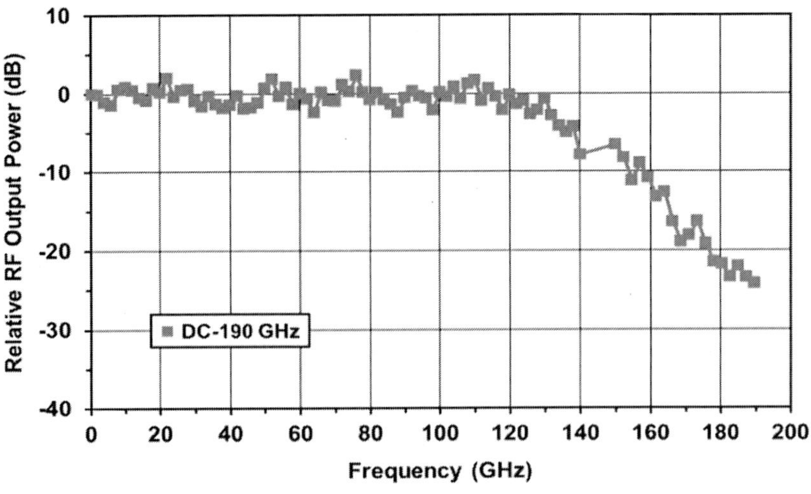

Figure 9. Measured relative RF output power from DC up to 190 GHz.

4.3. RF Output Power Measurements

Furthermore, the saturation effects were investigated within the 3 dB bandwidth. For calibrated power measurements at the 3 dB cut-off frequency (132 GHz), the F-band (90–140 GHz) SBD including an integrated WR-08 waveguide has been used together with a level set attenuator as the saturation of the SBDs starts at an RF input power of about −20 dBm. The maximum attenuation of the attenuator was set to 20 dB. Thus, the detected RF output power was measured for different optical input-power levels. In Figure 10, the measured RF output power is plotted against the photocurrent for the bias voltage of −8 V.

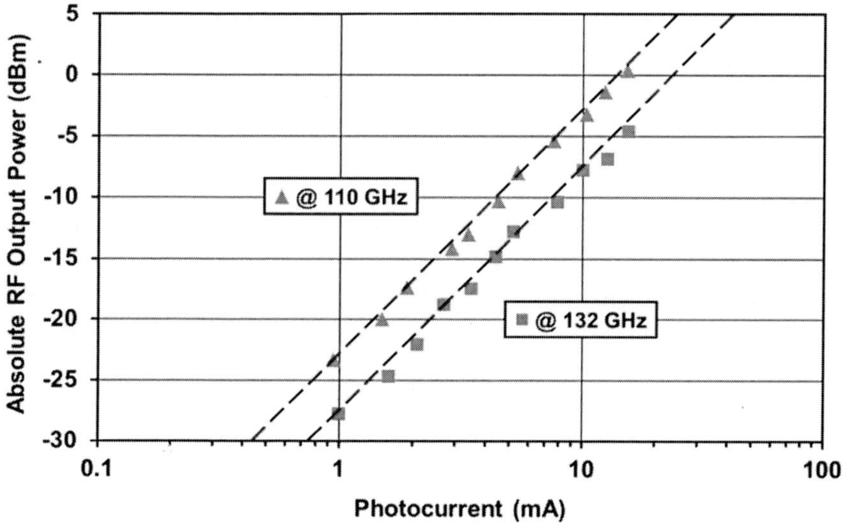

Figure 10. Measured RF output power plotted against the photocurrent for a bias voltage of −8 V at the frequencies of 110 GHz and 132 GHz. The dashed lines show the simulated output power for the RF responsivities of 0.24 A/W (110 GHz) and 0.14 A/W (132 GHz).

As shown in Figure 10, even at the 3 dB cut-off frequency no saturation effects occur and a maximum RF output power of −4.6 dBm is achieved, along with a photocurrent level of 15.5 mA. To identify the 1 dB compression point of the TTR-PD, a higher optical input-power level, which was not obtainable by the measurement setup, was required. However, the catastrophic optical damage of the passive optical waveguide facet caused by the applied lensed SMF has been calculated for an optical input-power level of about 24 dBm [24]. Maximum RF output-power levels exceeding 0 dBm (e.g., at 110 GHz) were attained as can be seen from Figure 10.

For hybrid integration of TTR-PDs on top of an optical silica waveguide, a limited number of modified TTR-PD chips have already been fabricated. Details will be reported elsewhere.

5. CONCLUSIONS

In this paper, we presented a novel high-speed InP-based waveguide triple transit region photodiode (TTR-PD) structure for hybrid integration with passive optical silica waveguides, which are fundamental to form an optical beam-steering network. As an application for future 5G mobile systems, we proposed a scheme for a hybrid-integrated InP/silicon-based photonic millimeter wave transmitter, enabling photonic RF beam forming and beam steering. For this approach, the TTR-PD chip was integrated on top of a passive optical silica waveguide to achieve evanescent optical coupling. Numerical beam propagation method analyses revealed that a coupling efficiency of around 90% is attained between a passive silica waveguide and the active InP-based waveguide TTR-PD, as well as an absorption of about 50% within a simulated TTR-PD length of 30 μm. For fabricated TTR-PD chips including a monolithically-integrated passive optical waveguide, a reasonable polarization dependent loss of less than 0.9 dB was achieved within the optical C-band (e.g., 0.73 dB at 1.55 μm). Moreover, a 3 dB cut-off frequency of 132 GHz was experimentally demonstrated. Within the 3 dB bandwidth, RF output-power levels higher than 0 dBm were measured for the fabricated TTR-PDs. The DC responsivity and the RF responsivity were found to be on the order of 0.52 A/W and 0.24 A/W, respectively.

ACKNOWLEDGMENTS

This work was supported by the European STReP IPHOBAC-NG (www.iphobac-ng.eu) under the grant No. 619870.

AUTHOR CONTRIBUTIONS

V.R., S.D., and A.S. developed and designed the component. V.R., P.L., and S.D. performed the simulations. V.R. developed the fabrication process and fabricated the devices. V.R., S.D., and B.K. performed the experimental characterization. V.R., B.K., and A.S. wrote the paper, supported by all other coauthors.

CONFLICTS OF INTEREST

The authors declare no conflict of interest.

REFERENCES

1. Nagatsuma, T.; Ito, H.; Ishibashi, T. High-power RF photodiodes and their applications. *Laser Photonics Rev.* **2009**, *3*, 123–137.

2. Ahmed, S.S.; Schießl, A.; Schmidt, L. A novel fully electronic active real-time imager based on a planar multistatic sparse array. *IEEE Trans. Microw. Theory Tech.* **2011**, *59*, 3567–3576.

3. Tseng, T.-F.; Wun, J.-M.; Chen, W.; Peng, S.-W.; Shi, J.-W.; Sun, C.-K. High-depth-resolution 3-dimensional radar-imaging system based on a few-cycle W-band photonic millimeter-wave pulse generator. *Opt. Express* **2013**, *21*, 14109–14119.

4. Stöhr, A. Photonic millimeter-wave generation and its applications in high data rate wireless access. In Proceedings of the IEEE International Topical Meeting on Microwave Photonics, Montreal, QC, Canada, 5–9 October 2010; pp. 7–10.

5. Seeds, A.J.; Shams, H.; Fice, M.J.; Renaud, C.C. Terahertz photonics for wireless communications. *J. Lightwave Technol.* **2015**, *33*, 579–587.

6. METIS. Available online: https://www.metis2020.com/ (accessed on 19 November 2015).

7. RAPID5G. Available online: http://www.rapid-5g.eu/ (accessed on 19 November 2015).

8. Stöhr, A.; Babiel, S.; Cannard, P.J.; Charbonnier, B.; van Dijk, F.; Fedderwitz, S.; Moodie, D.; Pavlovic, L.; Ponnampalam, L.; Renaud, C.C.; *et al.* Millimeter-wave photonic components for broadband wireless systems. *IEEE Trans. Microw. Theory Tech.* **2010**, *58*, 3071–3082.

9. ALMA. Available online: http://www.almaobservatory.org/ (accessed on 19 November 2015).

10. König, S.; Lopez-Diaz, D.; Antes, J.; Boes, F.; Henneberger, R.; Leuther, A.; Tessmann, A.; Schmogrow, R.; Hillerkuss, D.; Palmer, R.; *et al.* Wireless sub-THz communication system with high data rate. *Nat. Photonics* **2013**, *7*, 977–981.

11. Khani, B.; Rymanov, V.; Steeg, M.; Buck, A.; Dülme, S.; Stöhr, A. Compact E-band (71–86 GHz) bias-tee module for external biasing of millimeter wave photodiodes. In Proceedings of the IEEE International Topical Meeting on Microwave Photonics, Paphos, Cyprus, 26–29 October 2015.

12. Wun, J.-M.; Liu, H.-Y.; Lai, C.-H.; Chen, Y.-S.; Yang, S.-D.; Pan, C.-L.; Bowers, J.E.; Huang, C.-B.; Shi, J.-W. Photonic high-power 160-GHz signal generation by using ultrafast photodiode and a high-repetition-rate femtosecond optical pulse train generator. *IEEE J. Sel. Top. Quantum Electron.* **2014**, *20*, 10–16.

13. Urick, V.J.; McKinney, J.D.; Williams, K.J. High current photodetectors. In *Fundamentals of Microwave Photonics*, 1st ed.; John Wiley & Sons, Inc.: Hoboken, NJ, USA, 2015; pp. 351–382.

14. Preu, S.; Döhler, G.H.; Malzer, S.; Stöhr, A.; Rymanov, V.; Göbel, T.; Brown, E.R.; Feiginov, M.; Gonzalo, R.; Beruete, M.; *et al.* Principles of THz generation. In *Semiconductor Terahertz Technology: Devices and Systems at Room Temperature Operation*, 1st ed.; Carpintero, G., García Muñoz, L.E., Hartnagel, H.L., Preu, S., Räisänen, A.V., Eds.; John Wiley & Sons, Ltd.: Chichester, UK, 2015; pp. 3–68.

15. Rouvalis, E.; Chtioui, M.; Tran, M.; Lelarge, F.; van Dijk, F.; Fice, M.J.; Renaud, C.C.; Carpintero, G.; Seeds, A.J. High-speed photodiodes for InP-based photonic integrated circuits. *Opt. Express* **2012**, *20*, 9172–9177.

16. Beling, A.; Cross, A.S.; Piels, M.; Peters, J.; Fu, Y.; Zhou, Q.; Bowers, J.E.; Campbell, J.C. High-power high-speed waveguide photodiodes and photodiode arrays heterogeneously integrated on silicon-on-insulator. In Proceedings of the Optical Fiber Communication Conference and Exposition and the National Fiber Optic Engineers Conference, Anaheim, CA, USA, 17–21 March 2013; pp. 1–3.

17. El-Batawy, Y.M.; Youssef, S. Biasing effects on the characteristics and the optimization of mushroom waveguide photodetectors (WGPDs). In Proceedings of the International Symposium on Photonics and Optoelectronics (SPIE 9233), Suzhou, China, 22–24 August 2014.

18. Rymanov, V.; Stöhr, A.; Dülme, S.; Tekin, T. Triple transit region photodiodes (TTR-PDs) providing high millimeter wave output power. *Opt. Express* **2014**, *22*, 7550–7558.

19. Meijerink, A.; Roeloffzen, C.G.H.; Meijerink, R.; Zhuang, L.; Marpaung, D.A.I.; Bentum, M.J.; Burla, M.; Verpoorte, J.; Jorna, P.; Hulzinga, A.; *et al.* Novel ring resonator-based integrated photonic beamformer for broadband phased array receive antennas—Part I: Design and performance analysis. *J. Lightwave Technol.* **2010**, *28*, 3–18.

20. Zhuang, L.; Roeloffzen, C.G.H.; Meijerink, A.; Burla, M.; Marpaung, D.A.I.; Leinse, A.; Hoekman, M.; Heideman, R.G.; van Etten, W. Novel ring resonator-based integrated photonic beamformer for broadband phased array receive antennas—Part II: Experimental prototype. *J. Lightwave Technol.* **2010**, *28*, 19–31.

21. Wörhoff, K.; Heideman, R.G.; Leinse, A.; Hoekman, M. TriPleX: A versatile dielectric photonic platform. *Adv. Opt. Technol.* **2015**, *4*, 189–207.

22. Babiel, S.; Rymanov, V.; Stöhr, A. Hybrid integrated photonic millimeter wave emitter. In Proceedings of the Asia-Pacific Microwave Photonics Conference, Kyoto, Japan, 25–27 April 2012.

23. Rymanov, V.; Dülme, S.; Babiel, S.; Stöhr, A. Compact photonic millimeter wave (200–300 GHz) transmitters based on semicircular bow-tie antenna-integrated 1.55 μm triple transit region photodiodes. In Proceedings of the German Microwave Conference, Aachen, Germany, 10–12 March 2014; pp. 1–4.

24. Rymanov, V.; Khani, B.; Dülme, S.; Steeg, M.; Stöhr, A. Compact triple transit region photodiode module with WR-12 rectangular

waveguide output. In Proceedings of the IEEE International Topical Meeting on Microwave Photonics, Paphos, Cyprus, 26–29 October 2015.

6

P-CU$_2$O-SHELL/N-TIO$_2$-NANOWIRE-CORE HETEROSTUCTURE PHOTODIODES

Tsung-Ying Tsai†, Shoou-Jinn Chang†, Ting-Jen Hsueh† Han-Ting Hsueh†, Wen-Yin Weng†, Cheng-Liang Hsu† and Bau-Tong Dai†

ABSTRACT

This study reports the deposition of cuprous oxide [Cu_2O] onto titanium dioxide [TiO_2] nanowires [NWs] prepared on TiO_2/glass templates. The average length and average diameter of these thermally oxidized and evaporated TiO_2 NWs are 0.1 to 0.4 μm and 30 to 100 nm, respectively. The deposited Cu_2O fills gaps between the TiO_2 NWs with good step coverage to form nanoshells surrounding the TiO_2 cores. The p-Cu_2O/n-TiO_2 NW heterostructure exhibits a rectifying behavior with a sharp turn-on at approximately 0.9 V. Furthermore, the fabricated p-Cu_2O-shell/n-TiO_2-nanowire-core photodiodes exhibit reasonably large photocurrent-to-dark-current contrast ratios and fast responses.

Tsung-Ying Tsai, Shoou-Jinn Chang, Ting-Jen Hsueh, Han-Ting Hsueh, Wen-Yin Weng, Cheng-Liang Hsu and Bau-Tong Dai contributed equally to this work.

1. INTRODUCTION

UV photodetectors are important devices that have a range of commercial, research, and military applications. They can be used for space communication, ozone layer monitoring, and flame detection [1]. In recent years, high-performance GaN-based (including AlGaN and AlInN) [2–5], ZnO-based [6], and ZnSe-based [7] photodetectors have all been demonstrated. However, high-quality GaN-based UV photodetectors could only be prepared on a sapphire substrate, which is much more expensive as compared with a glass substrate. On the other hand, the photocurrent-to-dark-current contrast ratio of ZnO-based UV photodetectors is still low. Titanium dioxide [TiO_2] is a potentially useful wide direct-bandgap material (3.2 eV for anatase and 3.0 eV for rutile) for UV photodetectors, solar cells, and gas sensors due to its outstanding physical, chemical, and optical properties [8–10]. TiO_2 is a nontoxic naturally n-type semiconductor material which has a high-temperature stability and low-production costs.

For two-dimensional [2D] films, TiO_2 UV photodetectors such as metal-semiconductor-metal detectors and Schottky barrier diodes have been demonstrated [11, 12]. It is difficult to produce p- and n-type materials simultaneously, which is necessary for certain device applications. Zhang et al. reported the formation of a 2D TiO_2/Cu_2O composite film for a photocatalyst application using the metal ion-implantation method [13–15]. Cuprous oxide [Cu_2O] is naturally a p-type direct-bandgap semiconductor with a cubic crystal structure and a room-temperature bandgap energy of 2.17 eV [16], which makes it ideal for TiO_2-based p-n heterojunctions. Cu_2O can be deposited using methods such as thermal oxidation, anodic oxidation, sputtering, solution growth, sol-gel, and electro-deposition [17–24]. Among these methods, sputtering is commonly used in the semiconductor industry. By carefully controlling the growth parameters, high-quality 2D Cu_2O films can be produced by direct-current [DC] sputtering [18].

Recently, one-dimensional oxide semiconducting materials have attracted a lot of attention for potential application in optoelectronic devices due to their large surface-area-to-volume ratio [25]. Wu et al. reported the growth of TiO_2 nanowires [NWs] on glass substrates by the thermal oxidation-evaporation method [26, 27]. They produced single-crystalline TiO_2 NWs, whose size and density were controlled by adjusting the growth parameters. However, no report on the fabrication of p-Cu_2O-shell/n-TiO_2-nanowire-core heterojunction UV photodetectors could be found in the literature, to our knowledge. The present study reports the deposition of p-Cu_2O film onto n-TiO_2 NWs by DC sputtering and the fabrication of radial p-Cu_2O-shell/n-TiO_2-nanowire-core photodiodes. The physical, electrical, and optical properties of the fabricated radial p-Cu_2O-shell/n-TiO_2-nanowire-core photodiodes are discussed.

2. EXPERIMENTAL SECTION

Before the growth of TiO_2 NWs, a Corning 1737 glass substrate (Corning Display Technologies Taiwan Co., Ltd., Taipei City, Taiwan) was wet-cleaned with acetone and deionized water. The glass substrate was subsequently baked at 100°C for 10 min to evacuate moisture. A 400-nm-thick titanium [Ti] film layer was then deposited onto the glass substrate by electron-beam evaporation. Finally, the samples were annealed in a furnace at 700°C for 3 h to synthesize TiO_2 NWs in argon [Ar] ambiance. The crystal quality of the as-grown NWs was then characterized by an X-ray diffractometer [XRD] (MXP 18, MAC Science Co., Tokyo, Japan). The surface morphology of the samples and the size distribution of the NWs were characterized by a field-emission scanning electron microscope [FE-SEM] (JEOL JSM-7000F, JEOL Ltd., Tokyo, Japan).

To investigate the deposition of Cu_2O, glass was used as the substrate. The target used to deposit Cu_2O was a 4-N pure copper block mounted on the cathode. The distance between the target and the sample was fixed at 60 mm. A rotating magnet fixed on the backside of the cathode was used to enhance the plasma bombardment effect. During sputtering, the Ar flow rate, deposition time, base pressure, and chamber pressure were kept at 15 sccm, 10 min, 2×10^{-6} Torr, and 6 mTorr, respectively, and the DC power, O_2 flow rate, and substrate temperature were 200 W, 4 sccm, and 25°C, respectively. The crystallography and structure of the deposited Cu_2O and the Cu_2O/TiO_2 NWs were evaluated by XRD and FE-SEM.

Prior to the fabrication of p-Cu_2O-shell/n-TiO_2-nanowire-core photodiodes, a small piece of glass was used to cover the TiO_2 NWs to prevent the deposition of Cu_2O in these regions. A 200-nm-thick Cu_2O layer was subsequently deposited onto the TiO_2 NWs. A 500-nm-thick silver layer was then sputtered onto the Cu_2O layer and TiO_2 NWs to serve as the p-electrode and n-electrode with a shadow mask. Figure 1 schematically shows the structure of the fabricated p-Cu_2O-shell/n-TiO_2-nanowire-core photodiodes. A picoammeter (HP-4145B semiconductor parameter analyzer, Agilent Technologies, Sta. Clara, CA, USA), connected via a GPIB controller to a computer, was then used to measure the current-voltage [I-V] characteristics of the fabricated diodes under darkness. The photo responses of the devices were also measured. During photo-response measurements, a 4-W mercury vapor lamp emitting at 365 nm was used as the excitation source.

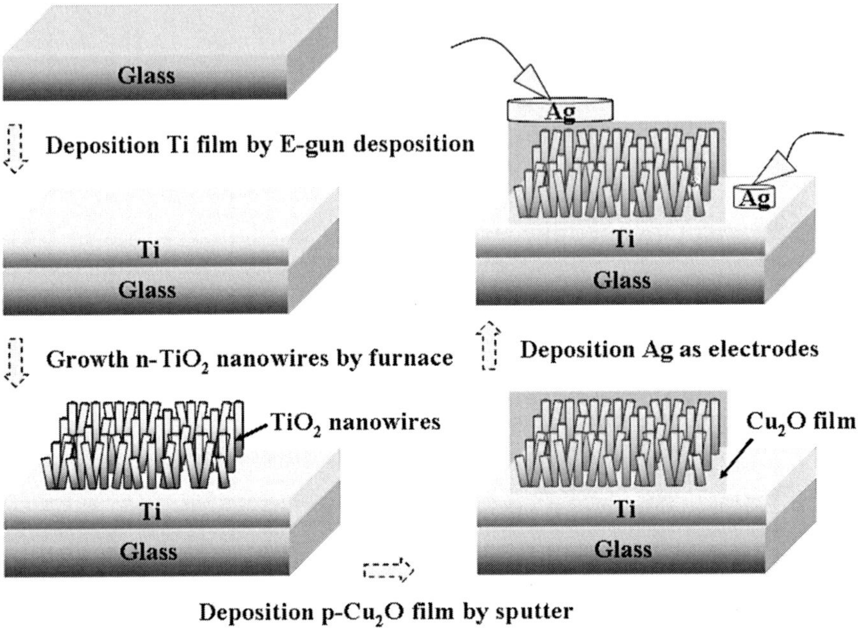

Figure 1 Schematic diagram of fabricated p-Cu_2O-shell/n-TiO_2-nanowire-core for photodiode measurements.

3. RESULTS AND DISCUSSION

Figure 2a shows a cross-sectional FE-SEM image of the TiO_2 NWs prepared on a Ti/glass template. It can be clearly seen that high-density TiO_2 NWs of various lengths were grown on the Ti/glass template. As shown in Figure 2a, it can be seen that the average length, diameter, and density of these TiO_2 NWs were 0.3 μm, 50 nm, and 60 wires/μm^2, respectively. Figure 2b shows a cross-sectional FE-SEM image of the sample with Cu_2O deposited on TiO_2 NWs. As shown, the deposited Cu_2O filled the gaps between the TiO_2 NWs with good step coverage to form radial Cu_2O/TiO_2 NWs. It was also found that the deposited Cu_2O formed at the sample surface after filling the gaps. In order to investigate the coating performance of Cu_2O, the deposited sample was scraped with tweezers into an alcohol solution, which was then ultrasonically treated for 20 min to disperse the NWs. The solution was dropped on carbon tape which was then placed on a hot plate to evacuate the alcohol. Figure 3a shows a SEM image of a single NW. Figures 3b and 3c show energy-dispersive X-ray [EDX] spectroscopic mapping images of Cu and

Ti, respectively. These figures correspond to the SEM image shown in Figure 3a. After the deposition of Cu_2O, Cu and Ti atoms were distributed over the entire NW. These results suggest that the sputtered Cu_2O not only forms the head portion of the nanoclubs, but also forms nanoshells surrounding the TiO_2 cores in the nanowire portion of the nanoclubs. The formation of such p-Cu_2O-shell/n-TiO_2-nanowire-core heterostructure should be able to provide us with a large junction area, which is important for the application of photodetectors.

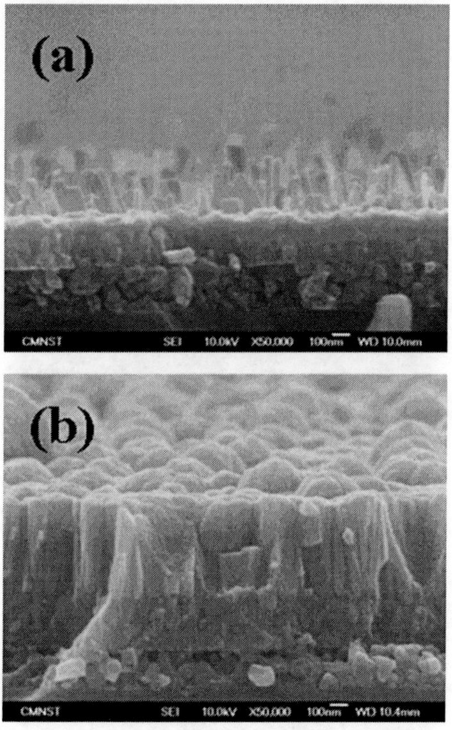

Figure 2 Cross-sectional FE-SEM images. (**a**) Pure TiO_2 nanowires and (**b**) the sample with Cu_2O deposited on TiO_2 nanowires at 300°C with 200 W DC power, 6 mTorr chamber pressure, and 3 sccm O_2 flow rate.

Figure 4 shows the crystallographic characteristics obtained from XRD measurements. For the pure TiO_2 NWs used for adhesion, the peaks were attributed to the rutile-TiO2 (110) phase (JCPDS Card No. 88-1175). For the p-Cu_2O/n-TiO_2 NWs, the peaks were attributed to the

(110) and (111) phases of the Cu_2O phase (JCPDS Card No. 78-2076). No Ti-related signal was found, indicating that the Ti film changed into a TiO_2 film after the annealing process.

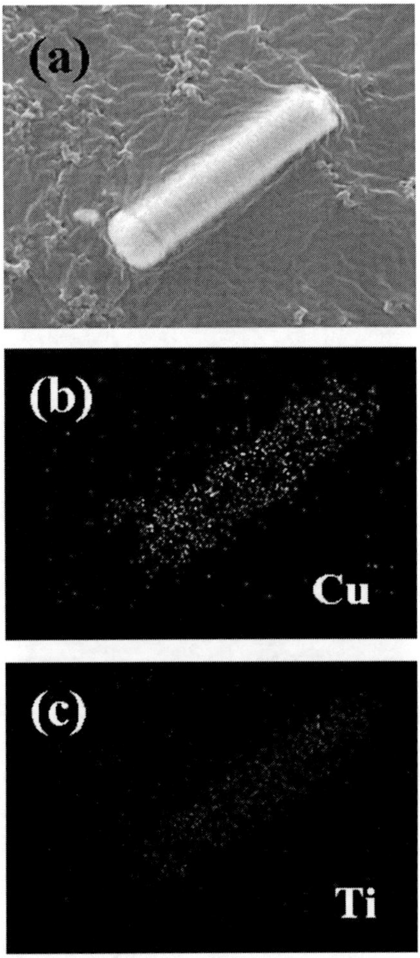

Figure 3 FE-SEM and EDX images. (**a**) An FE-SEM image of a single radial nanowire. EDX spectroscopic mapping images of (**b**) Cu and (**c**) Ti which were corresponded to (a).

Figure 5 shows the dark I-V characteristics measured from the fabricated radial Cu_2O/TiO_2 NWs. The rectifying behavior indicates that a p-n junction formed in the Cu_2O/TiO_2 NWs. The operation of the photodiode detector involves three steps (1) the generation of electron-hole [*e-h*] pairs by the absorption of incident light, whose photon energy exceeds the bandgap of the materials in the device; (2) the separation and transport of the *e-h* pairs by the internal electric field; and (3) the interaction of current with the external circuit to generate an output signal. Hence, the I-V characteristics of a photodiode in a dark environment are similar to those of a normal rectifying diode. If the p-n junction does not form, the generated *e-h* pairs will exhibit an ohmic character in the I-V curve and change the resistance. When a photodiode with a p-n junction is illuminated with optical radiation, the I-V characteristics shift according to the photocurrent and reverse current. The measured current in the photodiode, I_m, is:

$$I_m = I_d - I_{ph}$$

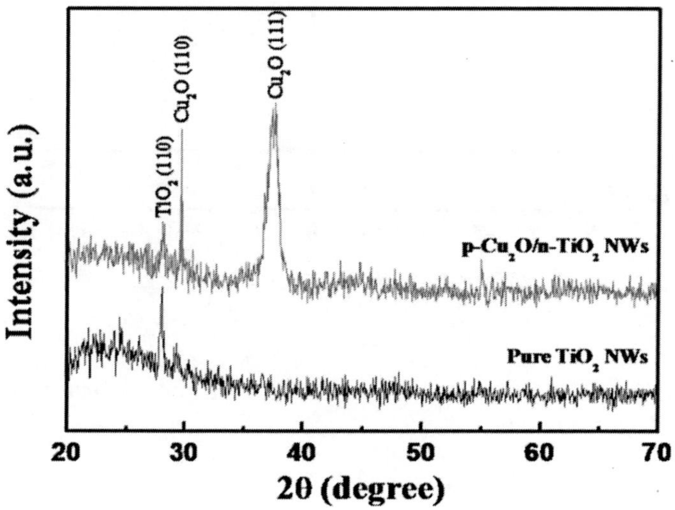

Figure 4 XRD measurements of pure TiO_2 nanowires and p-Cu_2O/n-TiO_2 nanowires obtained by DC sputtering.

where I_d is the dark current and I_{ph} is the photocurrent. The presence of a reverse current indicates that the photo response is due to the p-n junction, not the TiO_2 NWs or the Cu_2O. In the process of measurement under illumination, UV light passes through the TiO_2 and illuminates the array of the radial p-Cu_2O/n-TiO_2 NWs; *e-h* pairs are produced in the radial NWs when the energy of the UV light is absorbed. The *e-h* pairs are separated by the internal electric field, and a photocurrent is simultaneously generated. Under forward bias, the turn-on occurred at approximately 0.9 V. With a +5-V applied bias, the forward current of the device was 1.53×10^{-7} A, and with a -5-V applied bias, the reverse leakage current was 7.74×10^{-9} A.

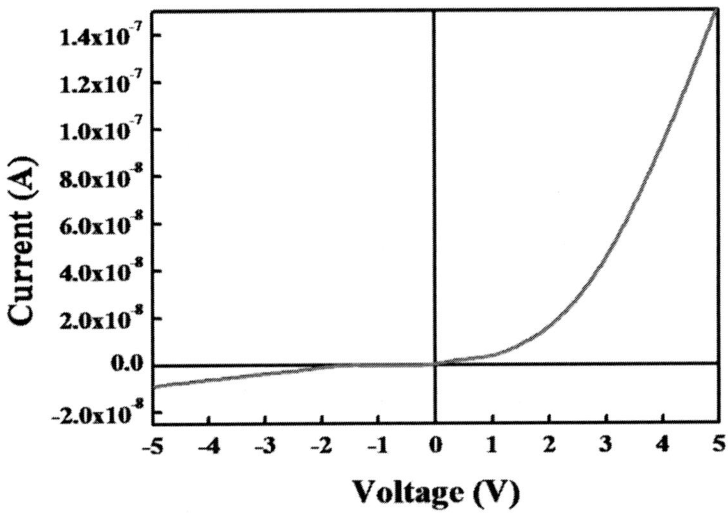

Figure 5 Dark I-V characteristic measured from the fabricated radial p-Cu_2O/n-TiO_2 nanowires.

Figure 6 shows the dynamic photo response measured from the fabricated p-Cu_2O-shell/n-TiO_2-nanowire-core photodiode. With a +10-V applied bias, the dark reverse leakage current of the diode was only around 3.37×10^{-8} A. However, the reverse leakage current increased rapidly to 1.15×10^{-6} A upon UV illumination. When the UV lamp was turned off, the reverse leakage current rapidly decreased to its original value. The reasonably large photocurrent-to-dark-current contrast ratio and the fast responses suggest that the radial p-Cu_2O-shell/n-TiO_2-nanowire-core photodiodes proposed in this study are potentially useful for UV detector applications.

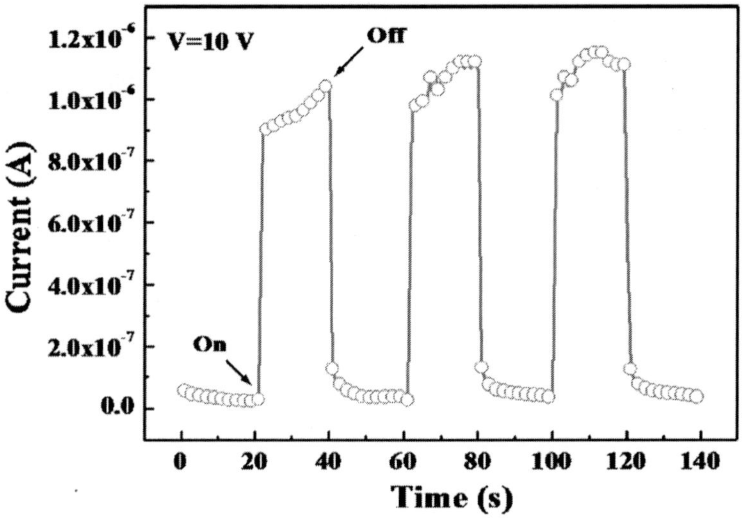

Figure 6 Dynamic photo response measured from the fabricated p-Cu_2O-shell/n-TiO_2-nanowire-core photodiode.

4. CONCLUSIONS

The deposition of Cu_2O onto well-aligned TiO_2 NWs by DC sputtering was reported. With the proper sputtering parameters, the deposited Cu_2O filled the gaps between the TiO_2 NWs with good step coverage to form radial p-Cu_2O/n-TiO_2 NWs that exhibited rectifying I-V characteristics. The fabricated radial p-Cu_2O-shell/n-TiO_2-nanowire-core photodiodes had a reasonably large photocurrent-to-dark-current contrast ratio and fast responses.

ACKNOWLEDGEMENTS

The authors would like to thank the National Science Council and Bureau of Energy, Ministry of Economic Affairs of Taiwan, Republic of China for the financial support under contract nos. 100-2221-E-006-040-MY2 and 100-D0204-6 and the LED Lighting Research Center of NCKU for the assistance on device characterization.

COMPETING INTERESTS

The authors declare that they have no competing interests.

AUTHORS' CONTRIBUTIONS

TYT carried out the nanowire experiments and data analysis and wrote the manuscript. SJC and TJH participated in data analysis and revised and finalized the manuscript. HTH designed the thin film and other experiments and data analysis. WYW participated in the revision of the manuscript. CLH provided the concept of the growth process of the nanowire. All the authors contributed to the preparation and revision of the manuscript and approved its final version.

REFERENCES

1. Monroy E, Calle F, Munoz E, Omnes F, Beaumont B, Gibart P: Visible-blindness in photoconductive and photovoltaic algan ultraviolet detectors. *J Electron Mater* 1999, 28: 240–245.

2. Chang SJ, Ko TK, Su YK, Chiou YZ, Chang CS, Shei SC, Sheu JK, Lai WC, Lin YC, Chen WS, Shen CF: GaN-based p-i-n sensors with ITO contacts. *IEEE Sens J* 2006, 6: 406–411.

3. Zhang J, Zhao H, Tansu N: Large optical gain AlGaN-delta-GaN quantum wells laser active regions in mid- and deep-ultraviolet spectral regimes. *Appl Phys Lett* 2011, 98: 171111.

4. Zhang J, Tong H, Liu G, Herbsommer JA, Huang G, Tansu N: Characterizations of Seebeck coefficients and thermoelectric figures of merit for AlInN alloys with various In-contents. *J Appl Phys* 2011, 109: 053706.

5. Zhang J, Kutlu S, Liu G, Tansu N: High-temperature characteristics of Seebeck coefficients for AlInN alloys grown by metalorganic vapor phase epitaxy. *J Appl Phys* 2011, 110: 043710.

6. Weng WY, Chang SJ, Hsu CL, Hsueh TJ: A ZnO nanowire phototransistor prepared on glass substrates. *ACS Appl Mat Interfaces* 2011, 3: 162–166.

7. Lin TK, Chang SJ, Su YK, Chiou YZ, Wang CK, Chang CM, Huang BR: ZnSe homoepitaxial MSM photodetectors with transparent ITO contact electrodes. *IEEE Trans Electron Devices* 2005, 52: 121–123.

8. Chiba Y, Islam A, Komiya R, Koide N, Han L: Conversion efficiency of 10.8% by a dye-sensitized solar cell using a TiO_2 electrode with high haze. *Appl Phys Lett* 2006, 88: 223505–223507.

9. Kopidakis N, Neale NR, Zhu K, Lagemaat JVD, Frank AJ: Spatial location of transport-limiting traps in TiO_2 nanoparticle films in dye-sensitized solar cells. *Appl Phys Lett* 2005, 87: 202106–202108.

10. Shen L, Zhu G, Guo W, Tao C, Zhang X, Liu C, Chen W, Ruan S, Zhong Z: Performance improvement of TiO_2 **/P3HT solar cells using CuPc as a sensitizer.** *Appl Phys Lett* 2008, 92: 073307–073309.

11. Xue H, Kong X, Liu Z, Liu C, Zhou J, Chen W, Ruan S, Xu Q: TiO_2 based metal-semiconductor-metal ultraviolet photodetectors. *Appl Phys Lett* 2007, 90: 201118–201120.

12. Kong X, Liu C, Dong W, Zhang X, Tao C, Shen L, Zhou J, Fei Y, Ruan S: Metal-semiconductor-metal TiO_2 ultraviolet detectors with Ni electrodes. *Appl Phys Lett* 2009, 94: 123502–123504.

13. Zhang KJ, Xu W, Li XJ, Zheng SJ, Xu G, Wang JH: Photocatalytic oxidation activity of titanium dioxide film enhanced by Mn non-uniform doping. *Trans Nonferrous Met SOC China* 2006, 16: 1069–1075.

14. Yasomanee JP, Bandara J: Multi-electron storage of photoenergy using Cu_2O-TiO_2 thin film photocatalyst. *Sol Energy Mat Sol Cells* 2008, 92: 348–352.

15. Zhang YG, Ma LL, Li JL, Yu Y: In situ Fenton reagent generated from TiO_2**/Cu**$_2$O composite film: a new way to utilize TiO_2 under visible light irradiation. *Environ Sci Technol* 2007, 41: 6264–6269.

16. Siripala W, Ivanovskaya A, Jaramillo TF, Baeck SH, McFarland EW: A Cu_2O/TiO_2 heterojunction thin film cathode for photoelectrocatalysis. *Sol Energy Mat Sol Cells* 2003, 77: 229–237.

17. Ghijsen J, Tjeng LH, Elp JV, Eskes H, Westerink J, Sawatzky GA, Czyzyk MT: Electronic structure of Cu_2O and CuO. *Phys Rev B* 1988, 38: 11322–11330.

18. Ishizuka S, Kato S, Okamoto Y, Sakurai T, Akimoto K, Fujiwara N, Kobayashi H: Passivation of defects in polycrystalline Cu_2O thin films by hydrogen or cyanide treatment. *Appl Surf Sci* 2003, 216: 94–97.

19. Herion J, Niekisch EA, Scharl G: Investigation of metal oxide/cuprous oxide heterojunction solar cells. *Sol Energy Mater* 1980, 4: 101–112.

20. Fortin E, Masson D: Photovoltaic effects in Cu_2O-Cu solar cells grown by anodic oxidation. *Solid State Electron* 1982, 25: 281–283.

21. Fernando CAN, Wetthasinghe SK: Investigation of photoelectrochemical characteristics of n-type Cu_2O films. *Sol Energy Mater Sol Cells*2000, 63: 299–308.

22. Armelao L, Barreca D, Bertapelle M, Bottaro Y, Sada C, Tondello E: A sol-gel approach to nanophasic copper oxide thin films. *Thin Solid Films* 2003, 442: 48–52.

23. Golden TD, Shumsky MG, Zhou Y, VanderWerf RA, Leeuwen RAV, Switzer JA: Electrochemical deposition of copper (I) oxide films. *Chem Mater* 1996, 8: 2499–2504.

24. Mahalingam T, Chitra JSP, Chu JP, Sebastian PJ: Preparation and microstructural studies of electrodeposited Cu_2O thin films. *Mater Lett*2004, 58: 1802–1807.

25. Hsueh TJ, Chen HY, Tsai TY, Weng WY, Yeh YM, Dai BT, Shieh JM: Si nanowire-based photovoltaic devices prepared at various temperatures. *IEEE Electron Dev Lett* 2010, 31: 1275–1277.

26. Wu JM, Shih HC, Wu WT: Electron field emission from single crystalline TiO_2 nanowires prepared by thermal evaporation. *Chen Phys Lett*2005, 413: 490–494.

27. Wu JM, Shih HC, Wu WT: Formation and photoluminescence of single-crystalline rutile TiO_2 nanowires synthesized by thermal evaporation. *Nanotechnology* 2006, 17: 105–109.

7

QUANTITATIVE AND CHEMICAL FINGERPRINT ANALYSIS FOR THE QUALITY EVALUATION OF*ISATIS INDIGOTICA* BASED ON ULTRA-PERFORMANCE LIQUID CHROMATOGRAPHY WITH PHOTODIODE ARRAY DETECTOR COMBINED WITH CHEMOMETRIC METHODS

Yan-Hong Shi [1], Zhi-Yong Xie [1], Rui Wang [1,*], Shan-Jun Huang [1], Yi-Ming Li [1] and Zheng-Tao Wang [2]

[1] School of Pharmacy, Shanghai University of Traditional Chinese Medicine, Shanghai 201203, China

[2] Institute of Chinese Materia Medica, Shanghai University of Traditional Chinese Medicine, Shanghai 201203, China

ABSTRACT

A simple and reliable method of ultra-performance liquid chromatography with photodiode array detector (UPLC-PDA) was developed to control the quality of *Radix Isatidis* (dried root of *Isatis indigotica*) for chemical fingerprint analysis and quantitative analysis of eight bioactive constituents, including *R,S*-goitrin, progoitrin, epiprogoitrin, gluconapin, adenosine, uridine, guanosine, and hypoxanthine. In quantitative analysis, the eight components showed good regression ($R > 0.9997$) within test ranges, and the recovery method ranged from 99.5% to 103.0%. The UPLC fingerprints of the *Radix Isatidis* samples were compared by performing chemometric procedures, including similarity analysis, hierarchical clustering analysis, and principal component analysis. The chemometric procedures classified *Radix Isatidis* and its finished products such that all samples could be successfully grouped according to crude herbs, prepared slices, and adulterant *Baphicacanthis cusiae* Rhizoma et Radix. The combination of quantitative and chromatographic fingerprint analysis can be used for the quality assessment of *Radix Isatidis* and its finished products.

KEYWORDS

Radix Isatidis; UPLC-PDA; quality control; quantitative; fingerprint

1. INTRODUCTION

Radix Isatidis is the dried root of *Isatis indigotica* Fort. (Fam. Cruciferae), which is known as *Banlangen* (BLG) or *Bei-Banlangen. Baphicacanthis cusiae* Rhizoma et Radix is the dried rhizome and root of *Baphicacanthus cusia* (Nees) Bremek. (Fam. Acanthaceae), which is known as *Nan-Banlangen* (NBLG). In several southern regions in China, NBLG has been improperly used as *Radix Isatidis* even though both have been officially listed in Chinese Pharmacopoeia (State Pharmacopoeia 2000) as two different crude herbs [1–4].

Radix Isatidis and its finished products have important functions in preventing and treating influenza, tonsillitis, and malignant infectious diseases [5–7], especially severe acute respiratory syndrome (SARS) and H1N1-influenza [8–11] because of its anti-viral, anti-bacterial, anti-inflammatory, anti-tumor, and immune regulatory functions. Moreover, numerous gratifying successes of the *Radix Isatidis* anti-viral effect have been reported [12–15]. *Radix Isatidis* has become an important component in various traditional Chinese medicine preparations (TCMPs), of which Banlangen Granules (BLGG) is widely used for toxic-heat removal and as an anti-viral drug in clinical practice.

The chemical constituents in *Radix Isatidis* are as follows: nucleosides, glucosinolates, amino acids, polysaccharides, alkaloids, organic acid, trace elements, and so on [16–18]. Compared with other *Radix Isatidis* extraction methods, the water extraction method has the most significant anti-viral, anti-bacterial, and anti-endotoxic effects, and the chemical constituents of glucosinolates (*R,S*-gotrin, progoitrin, epiproguotrin, and gluconapin) and nucleosides (hypoxanthine, adenosine, uridine, and guanosine) are the major bioactive components [19–21]. Therefore, using the non-polar components of Indigotin, Indirubin, and the nonspecific amino acid constituent as quality control markers of *Radix Isatidis* is unsuitable [22–27].

Glucosinolates are one of the characteristic components of Cruciferae. They degrade under endogenous myrosinase effects to work on multi-bioactivities including anti-virus and anti-bacteria [28,29]. *R,S*-goitrin reflects bioactivities that are relevant to the *Radix Isatidis* effects with high specificity. In our previous studies, we established thin layer chromatography (TLC) identification and high-performance liquid chromatography (HPLC) assay methods for the quality control of *Radix Isatidis*using *R,S*-goitrin as the marker [30], and these methods have been adopted by the ChP 2010 Edition [1]. Progoitrin, epiproguotrin, and *R,S*-gotrin all exist in *Radix Isatidis*. Under the effects of myrosinase, some parts or all of progoitrin and epiproguotrin transfer to *R,S*-goitrin via degradation [31].

In previous studies, based on the issues related to *Radix Isatidis* such as the improper use of NBLG, glucosinolate degradation, lack of specificity using amino acid as the BLGG quality control marker, establishing proper and scientific-based methods for quality control of *Radix Isatidis* and its finished products is necessary. Current studies about quantitative and chemical fingerprinting of *Radix Isatidis* depend on non-polar extracts or spots of chemical compositions [32–35]. Chemical fingerprint and quantitative analysis has become one of the most frequently applied approaches in the quality control of traditional Chinese medicine (TCM) and its finished products [36–41]. Studies about combining chromatographic fingerprint and multi-ingredient quantification by ultra-performance liquid chromatography with photodiode array detector (UPLC-PDA) for the quality control of *Radix Isatidis* and its preparations have not been reported.

This study aims to reveal the correlation and consistency of quality control in crude herbs, prepared slices, and *Radix Isatidis* preparations. A simple, accurate, and practical UPLC-PDA method was developed for the simultaneous determination of eight bioactive components in *Radix*

Isatidis and its preparations. The chemical fingerprints of *Radix Isatidis*from various sources were established and investigated by similarity analysis (SA), hierarchical clustering analysis (HCA), and principal component analysis (PCA). The combination of chromatographic fingerprint analysis and the simultaneous determination of the eight bioactive components offer a more comprehensive strategy for the quality evaluation of *Radix Isatidis* and its finished products.

2. RESULTS AND DISCUSSION

2.1. Optimization of UPLC Conditions

Different UPLC parameters were examined and compared, including various columns, mobile phases, detection wavelengths, and gradient elution conditions to obtain as much chemical information as possible and to determine the best separation mechanism in chromatograms.

Four kinds of reversed-phase columns, namely, Waters ACQUITY UPLC BEH C_{18} (1.7 μm, 2.1 mm × 50 mm/100 mm) and Thermo Syncronis C_{18}/Waters ACQUITY UPLC HSS T_3 (1.7 μm, 2.1 mm × 100 mm) were investigated and compared. The Waters ACQUITY UPLC BEH C_{18} (1.7 μm, 2.1 mm × 100 mm) column had good peak separation and sharp peaks.

The effect of mobile phase composition (methanol-water and acetonitrile-water with different modifiers including acetic acid, formic acid, phosphoric acid, and triethylamine) on chromatographic separation was investigated. Adding 0.1% triethylamine in the mobile phase and adjusting the pH to 4.0 with formic acid provided a better resolution and separation of the eight bioactivity components, and resulted in high precision sensitivity and selectivity.

Based on the maximum absorption and full-scan experiment of the marker components in the UV spectra of the three-dimensional chromatograms obtained by PDA detection, the detection wavelengths at 210, 230, 254, 280 nm were selected to compare the peak number and peak resolution of all marker compounds. Finally, the wavelength was set at 254 nm (Figure 1).

2.2. Optimization of Extraction Methods

Satisfactory extraction efficiency was obtained by comparing water-refluxing, ultrasonic, and soxhlet extraction methods. Refluxing extraction was simpler and more effective for nucleosides and glucosinolates among the other methods. Therefore, this method was used in further experiments. In this study, different concentrations (0%, 20%, 50%, 75%, 90%, and 100%) of ethanol solutions, sample-solvent ratios (1:10, 1:20, 1:50, and 1:100, *w*/*v*), and extraction times (20, 30, 45, and 60 min) were used for the BLG extraction procedure (Batch No. 20101121). As a result, the best extraction condition was established as follows: the samples were extracted by refluxing extraction using 20 mL of water as the extraction solvent, and the duration was 30 min.

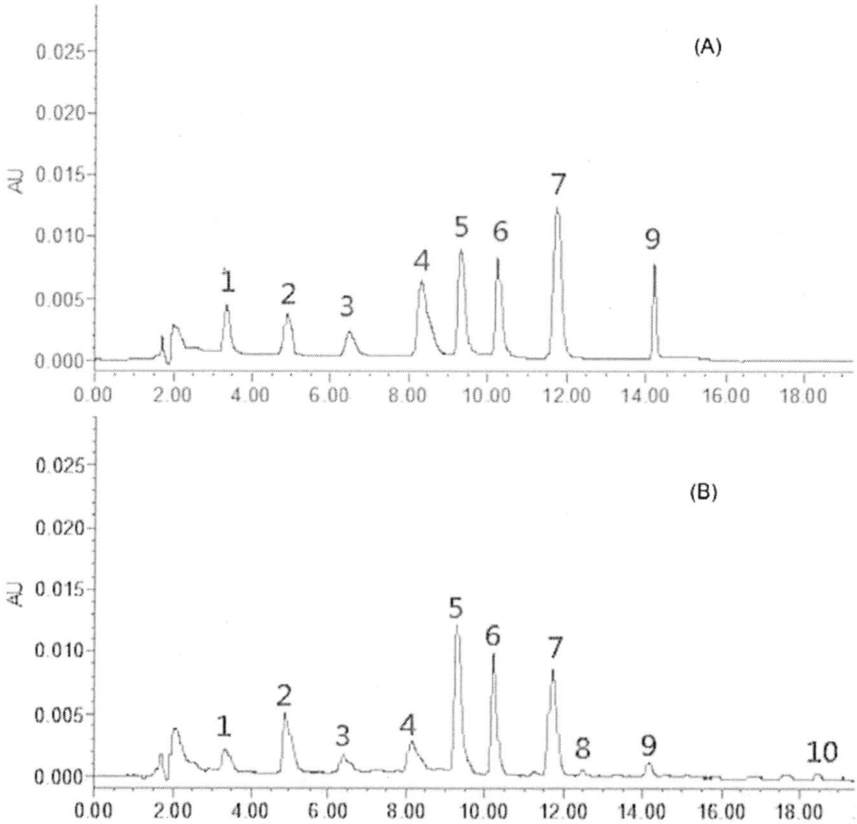

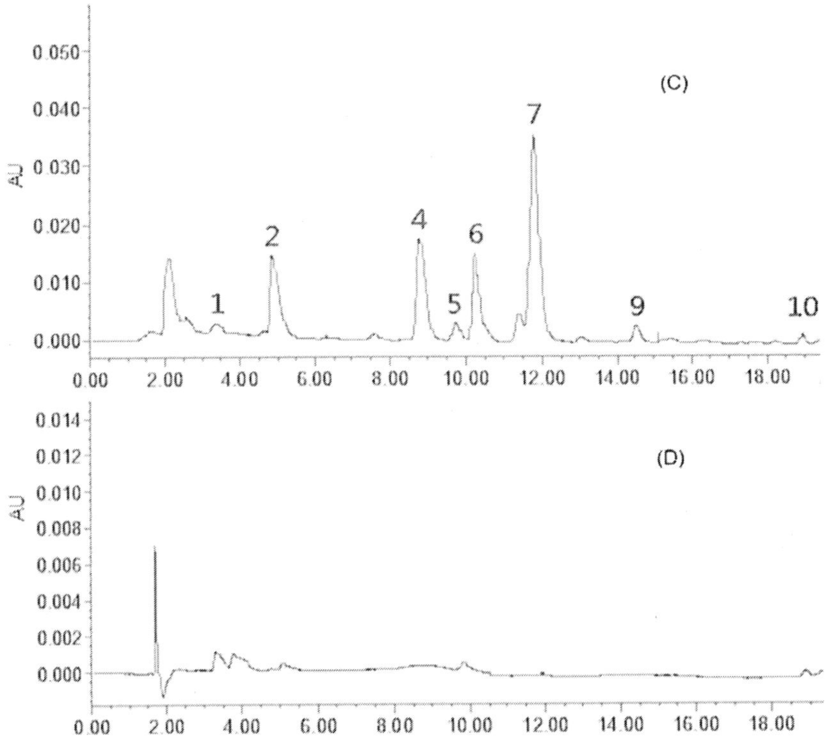

Figure 1. Typical chromatograms for determining the eight bioactive compounds in different samples. (**A**) Mixed standards; (**B**) *Radix Isatidis* (No. 20101121); (**C**) Banlangen Granules (BLGG) (No. 090921); (**D**) Nan-Banlangen (NBLG) (No. 060308-1). Peaks 1 = hypoxanthine, 2 = uridine, 3 = progoitrin, 4 = epiprogoitrin, 5 = adenosine, 6 = guanosine, 7 = *R,S*-goitrin, and 9 = luconapin.

2.3. Method Validation of Quantitative Analysis

The method was validated in terms of linearity, limit of detection (LOD), limit of quantification (LOQ), precision, reproducibility, stability, and recovery test.

2.3.1. Calibration Curves, LOD, and LOQ

Methanol stock solutions containing eight analytes were diluted to appropriate concentrations for calibration curve construction. The analyte solutions at six different concentrations were injected in triplicate, and the calibration curves were established by plotting the peak area (*Y*) versus the concentration (*x*) of each component. LOD and LOQ, which were expressed by 3-and 10-fold of the signal-to-noise ratio (*S/N*), were also determined. The detailed information regarding the calibration curves, linear ranges, LODs, and LOQs of the eight bioactive compounds are listed in Table 1.

Table 1. Regression data, limits of detection (LODs), and limits of quantification (LOQs) for the eight bioactive constituents.

Compound	Regression equation ($Y = ax + b$) [a]	R [b]	Linear range ($\mu g\ mL^{-1}$)	LOD [a] ($\mu g\ mL^{-1}$)	LOQ [b] ($\mu g\ mL^{-1}$)
Hypoxanthine	$Y = 914.32x - 684.64$	0.9999	6.50–130.00	0.015	0.049
Uridine	$Y = 1945.9x + 263.11$	0.9999	4.12–82.40	0.012	0.041
Adenosine	$Y = 6870.2x - 3669.6$	0.9999	4.26–85.20	0.005	0.017
Guanosine	$Y = 12618x - 1461.7$	1.0000	3.90–78.00	0.017	0.045
Progoitrin	$Y = 2579.2x - 886.87$	0.9997	2.30–34.50	0.563	1.875
Epiprogoitrin	$Y = 4336.8x - 68.733$	0.9998	2.30–34.50	0.708	2.100
R,S-goitrin	$Y = 1756.6x - 757.89$	1.0000	4.50–90.00	0.014	0.045
Gluconapin	$Y = 1865.8x - 240.44$	0.9999	2.05–30.75	0.456	1.640

[a] *Y* and *x* stand for the peak area and the injection quantity (μg) of each standard substance, respectively;
[b] *R* = correlation coefficient, *n* = 6.

2.3.2. Precision, Reproducibility, Stability, and Recovery

As shown in Table 2, the precision based on the peak area measurements of the eight bioactive components were higher than 0.48% (RSD, *n* = 6, S-01). The reproducibility (RSD, *n* = 6, S-01) of the proposed method based on six replicate injections was in the range of 0.10% to 1.24%. The stability (RSD, *n* = 6, S-01) of the measurements over 3 days for the eight compounds was 0.22% to 1.44%. The recovery test was performed by the standard addition method. Low, medium, and large high amounts of the standards were added to the known sample (S-01). Extraction and analysis were performed as described in Section 2.4. The mean recovery was calculated according to the following formula: recovery (%) = (amount found–original amount)/amount spiked × 100%, and RSD (%) = (SD/mean) × 100%. The mean recovery of the eight bioactive compounds was 99.5%–103.0%, and the RSD value was 0.73%–1.81%.

Table 2. Precision, reproducibility, stability, and recovery of the eight bioactive constituents.

Compound	Precision RSD (%) (n = 6)	Reproducibility RSD (%) (n = 6)	Stability RSD (%) (n = 6)	Recovery (%) (n = 9) Mean ± RSD (%)
Hypoxanthine	0.15	0.46	0.22	101.5 ± 1.63
Uridine	0.18	0.10	0.39	101.2 ± 1.33
Adenosine	0.07	0.61	0.49	101.1 ± 1.11
Guanosine	0.12	1.24	1.30	100.9 ± 1.72
Progoitrin	0.11	0.19	0.24	99.5 ± 1.81
Epiprogoitrin	0.11	1.92	1.44	100.2 ± 1.73
R,S-goitrin	0.48	0.72	0.91	100.4 ± 0.73
Gluconapin	0.31	0.33	0.48	103.0 ± 1.14

2.4. Sample Analysis

The newly established analytical method was subsequently applied to determine simultaneously the eight bioactive components in 21 commercial samples of *Radix Isatidis* and its finished products from different provinces or manufacturers in China. All samples were analyzed using the optimized extraction method in optimized UPLC conditions. Each sample was analyzed in triplicate to determine the mean content ($mg \cdot g^{-1}$), and the results are tabulated in Table 3.

Glucosinolates are the characteristic constituents in the Cruciferae plants [28,29]. The results from the quantitative analysis (Table 3) showed that the crude herbs generally contained the eight selected constituents and the glucosinolate content ranged from 6.48 mg g^{-1} to 73.63 mg g^{-1}. Table 3 and Figure 2 show that the average contents of epiprogoitrin, progoitrin, and *R,S*-goitrin in *Radix Isatidis* crude herbs are 5.86, 6.25, and 0.07 mg g^{-1}. However, the average epiprogoitrin and progoitrin contents in prepared *Radix Isatidis* slices significantly decreased (2.53 and 2.68 mg g^{-1}, respectively), whereas the *R,S*-goitrin content clearly increased (0.64 mg g^{-1}). The traditional processing methods improve the glucosinolate biotransformation and increase the degradation product (*R,S*-goitrin) content [31].

The total content of eight compounds in different samples, especially those from different sources and harvesting times, were significantly different. Moreover, epiprogoitrin and progoitrin could not be distinctly detected (below LOD) in BLGG that was prepared and precipitated with ethanol by the water extracts of *Radix Isatidis* [1].

Table 3. The contents (mg g-1) of eight targets in 21 commercial samples (n = 3).

Name	No.[a]	Content [b] (mg g^{-1})								
		1 [c]	2	3	4	5	6	7	9	Total
Prepared slices	S-01	0.137	0.414	5.938	6.605	0.492	0.525	0.776	4.051	18.937
	S-02	0.022	0.279	1.529	1.500	0.293	0.377	0.899	0.870	5.768
	S-03	0.042	0.436	2.389	2.094	0.343	0.465	0.789	1.210	7.767
	S-04	0.024	0.283	2.465	2.293	0.342	0.336	0.414	1.190	7.347
	S-05	0.019	0.278	3.331	3.355	0.347	0.345	0.551	1.977	10.203
	S-06	0.037	0.416	1.178	1.413	0.372	0.424	0.732	0.401	4.974
	S-07	0.021	0.384	1.416	1.750	0.336	0.347	0.483	0.574	5.310
	S-08	0.336	0.345	1.950	2.441	0.348	0.379	0.436	0.501	6.735
Crude herbs	S-09	0.056	0.083	5.929	6.605	0.114	0.095	0.057	10.020	22.959
	S-10	0.022	0.057	7.345	6.565	0.127	0.132	0.083	9.458	23.790
	S-11	0.028	0.070	5.286	6.424	0.120	0.116	0.070	10.130	22.244
	S-12	0.035	0.084	3.934	5.660	0.124	0.137	0.123	7.235	17.333
	S-13	0.048	0.057	6.396	6.534	0.105	0.135	0.062	6.747	20.084
	S-14	0.040	0.059	5.507	6.253	0.122	0.327	0.084	9.611	22.003
	S-15	0.033	0.070	6.628	5.713	0.105	0.207	0.054	5.498	18.309
Granules	S-16	N.D. [d]	0.045	N.D.	N.D.	0.006	0.008	0.085	0.330	0.473
	S-17	0.004	0.022	N.D.	N.D.	N.D.	N.D.	0.027	0.094	0.147
	S-18	N.D.	0.058	N.D.	N.D.	0.003	0.005	0.075	0.576	0.716
	S-19	N.D.	0.021	N.D.	N.D.	0.001	N.D.	0.033	0.100	0.154
	S-20	0.006	0.061	N.D.	N.D.	0.007	0.006	0.087	0.515	0.681
	S-21	0.036	0.053	N.D.	0.102	0.012	0.010	0.067	0.169	0.448

[a] No. means sample numbers as listed in Table 4; [b] Values in mg g^{-1} of dry raw materials; and in mean ± SD, n = 3, the SD was < 4% of the mean, which is not shown for clarity; [c] (1–7,9) mean the bioactive constituents number as listed in Figure 1; [d] N.D.: below LOD.

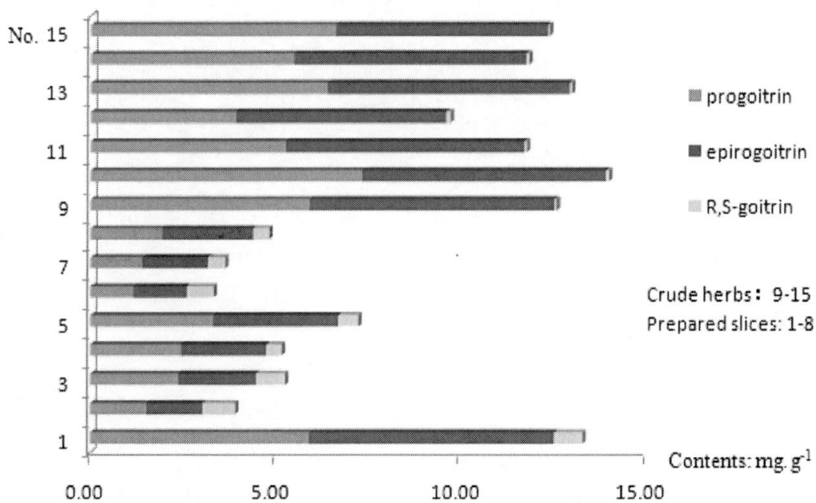

Figure 2. Bar graph of the contents of progoitrin, epiprogoitrin, and *R,S*-gotrin in 15 sample batches.

Overall, the internal qualities of 21 batches of *Radix Isatidis* samples from different sources with different geographical sources were varied, and the qualities needed evaluation by chemical fingerprinting.

2.5. UPLC Fingerprint of Radix Isatidis

Altogether, 15 batches of samples were analyzed, and all chromatograms were introduced into the Computer-Aided Similarity Evaluation System for Chromatographic Fingerprint of TCM (China Committee of Pharmacopeia, 2004). Peaks existing in all sample chromatograms were assigned as the "common peak," and 10 common peaks were observed between 3 and 20 min in all 15 batches (Figure 1). Eight common peaks (peak 1, 2, 3, 4, 5, 6, 7, and 9) were identified as hypoxanthine, uridine, progoitrin, epiprogoitrin, adenosine, guanosine, *R,S*-goitrin, and gluconapin with reference substances (Figure 3). Peak 7 (*R,S*-goitrin, RT = 11 min), which was one of the most important active constituents of *Radix Isatidis* (China Pharmacopoeia Committee 2010), was chosen as the internal reference peak to calculate the relative retention time (RRT) and relative peak area (RPA) of the other peaks.

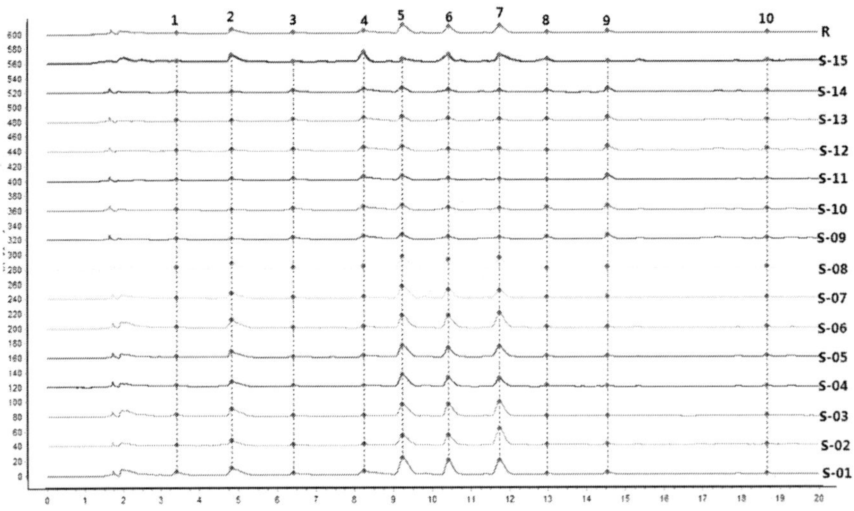

Figure 3. Ultra-performance liquid chromatography (UPLC) fingerprint chromatograms of 15 batches of *Radix Isatidis* (R: digital standard fingerprint).

Figure 1 shows that the investigated nucleosides, glucosinolates, and other compounds in *Radix Isatidis* were separated and determined using the developed UPLC-PDA method. The differences of the water-soluble bioactive components from BLG and NBLG were recognized, and the components could be rapidly and efficiently differentiated by the chromatographic method (Figure 1B,D). Moreover, the bioactive constituents of glucosinolates and nucleosides were simultaneously shown by the same chromatographic method, and the high correlation between BLG and BLGG (Figure 1B,C) provided a basis to investigate the relationship between *Radix Isatidis* and its preparations.

2.5.1. Similarity Analysis (SA)

The State Food and Drug Administration (SFDA) suggested that all herbal chromatograms should be evaluated in terms of similarity by calculating the correlation coefficient and/or angle cosine value of the original data [36–38]. Therefore, SA was conducted based on the standard fingerprints, and the results are shown in Table 4. The same samples showed similarities (crude herbs: 0.752 to 0.820; prepared slices: 0.933 to 0.991). The same samples had similar constituents with the slight difference resulting from the environmental conditions and the planting techniques. However, the crude herbs and the prepared slices were significantly different because of the production process and the contents of main bioactive constituents. The biotransformation, the glucosinolate contents, and the degradation products have the most important influence on the similarities.

2.5.2. Hierarchical Cluster Analysis (HCA)

The HCA results clearly showed that the samples were appropriately divided into two main clusters related to the *Radix Isatidis* type (Figure 4). Cluster-I was S01 to S08, which were the prepared slices of *Radix Isatidis*, whereas Cluster-II was S09 to S15, which were the crude herb samples. The HCA result was fully consistent with the actual situation of samples.

Table 4. All samples used in this work and their corresponding similarities.

Name	No.	Batch No.	Resource	Origins	Similarity
Prepared Slices	S-01	20101121	Anhui	*Isatis indigotica* Fort.	0.991
	S-02	101001	Anhui	*Isatis indigotica* Fort.	0.933
	S-03	8100182	Anhui	*Isatis indigotica* Fort.	0.948
	S-04	100709	Anhui	*Isatis indigotica* Fort.	0.947
	S-05	20110213	Jiangsu	*Isatis indigotica* Fort.	0.967
	S-06	100713	Mongolia	*Isatis indigotica* Fort.	0.954
	S-07	100602	Mongolia	*Isatis indigotica* Fort.	0.954
	S-08	101120	Zhejiang	*Isatis indigotica* Fort.	0.967
Crude herbs	S-09	20110224-1	Shanghai	*Isatis indigotica* Fort.	0.777
	S-10	20110224-2	Shanghai	*Isatis indigotica* Fort.	0.774
	S-11	20110224-3	Shanghai	*Isatis indigotica* Fort.	0.752
	S-12	20080902	Harbin	*Isatis indigotica* Fort.	0.759
	S-13	blg-081020	Henan	*Isatis indigotica* Fort.	0.818
	S-14	20110205	Anhui	*Isatis indigotica* Fort.	0.820
	S-15	20081017	Hebei	*Isatis indigotica* Fort.	0.811
Granule	S-16	090921	Shanghai	-	-
	S-17	100302	Shanghai	-	-
	S-18	A0F090	Guangzhou	-	-
	S-19	L9F043	Guangzhou	-	-
	S-20	100303	Sichuan	-	-
	S-21	100702	Sichuan	-	-
Nan-Banlangen	S-22	060308-1	Jiangxi	*Baphicacanthus eusia* (Nees) Bremek.	-
	S-23	1678-3	Guizhou	*Baphicacanthus eusia* (Nees) Bremek.	-
	S-24	20110923	Fujian	*Baphicacanthus eusia* (Nees) Bremek.	-
	S-25	20111025	Fujian	*Baphicacanthus eusia* (Nees) Bremek.	-

2.5.3. Principal Component Analysis (PCA)

PCA, which is an unsupervised multivariate data analysis approach, is appropriate when a function of many attributes is believed to be involved in different samples [40,41]. PCA was employed to analyze the relationship of the 15 *Radix Isatidis*samples (seven crude herbs and eight prepared slices) from different sources, and the score plot derived from PCA is shown in Figure 5A.

On the basis of eigenvalues > 1, the first two principal components, PC1 and PC2, are often used to provide a convenient visual aid for identifying inhomogeneity in the data sets. The samples were clustered into two main groups. The PCA loading plot (Figure 5B) indicated that peak 4 (progoitrin) and peak 7 (*R,S*-goitrin) showed the greatest influence on the scores. Peak 2 (uridine) and other compounds also affected the scores. Progoitrin content in the crude herbs was obviously higher than that in the prepared slices, whereas *R,S*-goitrin content had a reverse trend. The difference between crude herbs and prepared *Radix Isatidis* slices could be generally tracked to the transformation of glucosinolates to their degradation products.

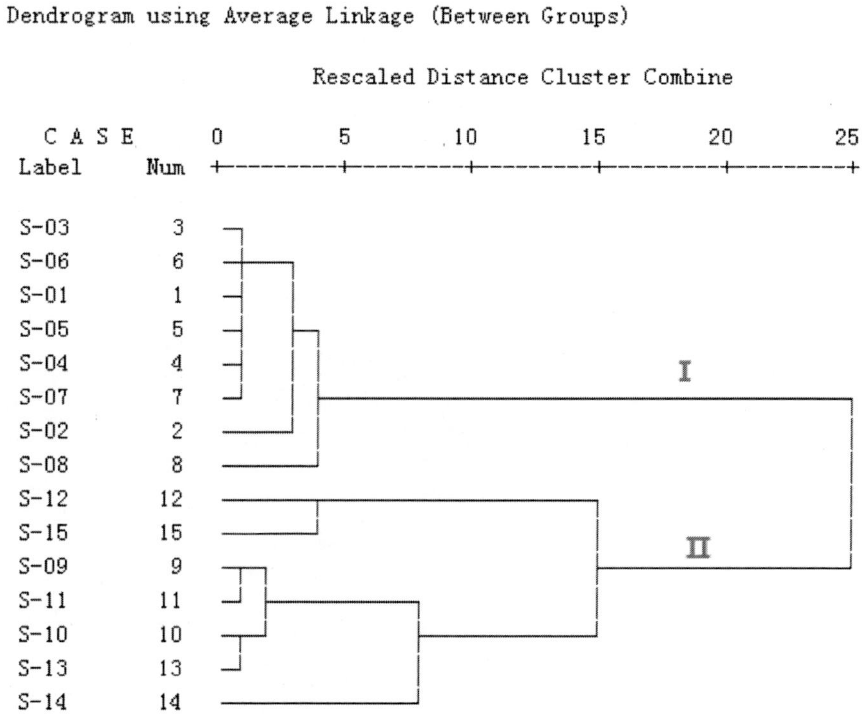

Figure 4. Dendrogram of hierarchical clustering analysis (HCA) for the 15 tested samples. The hierarchical clustering was done by the SPSS software (version 15.0; IBM: Chicago, IL, USA, 2006). Ward's method was performed, and Squared Euclidean distance was selected as a measurement.

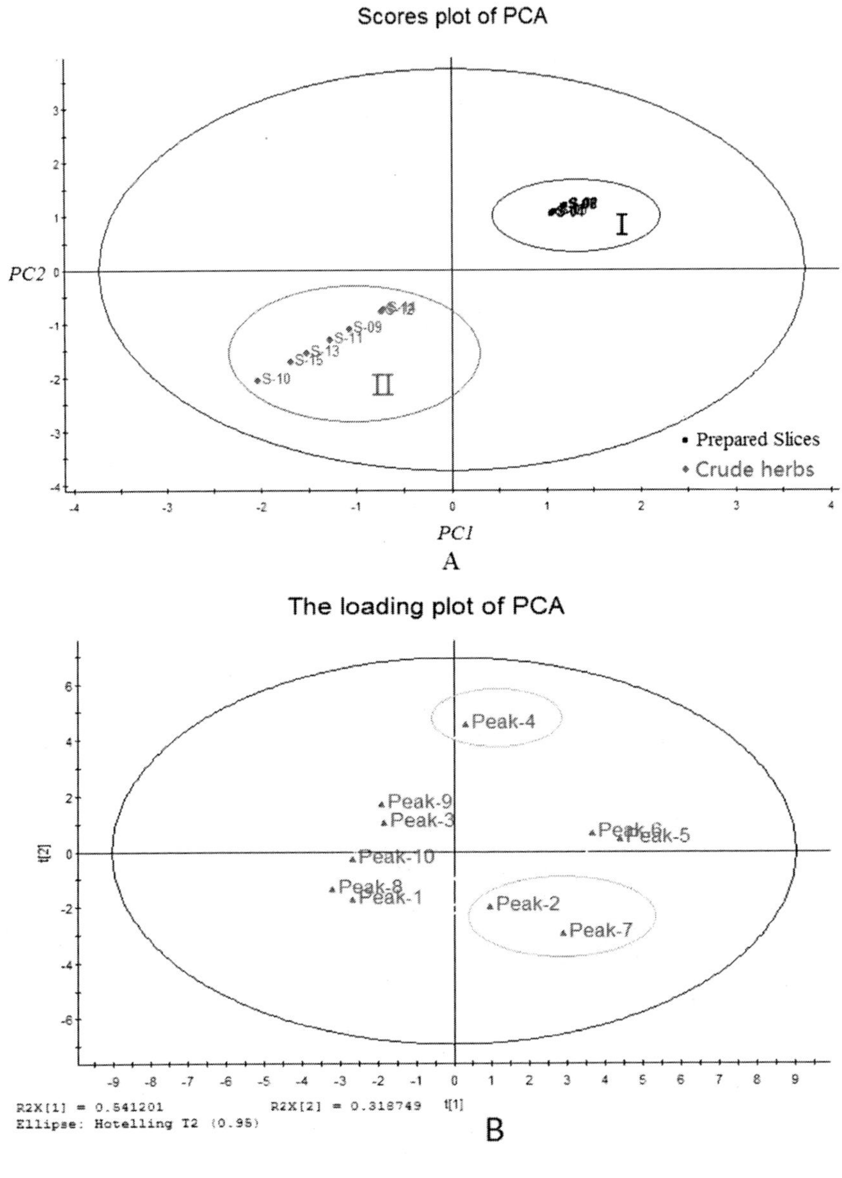

Figure 5. Score plot of principal component analysis (PCA) (**A**) and PCA loading plot (**B**) of *Radix Isatidis*.

3. EXPERIMENTAL SECTION

3.1. Materials, Reagents, and Chemicals

Eight BLG (prepared slices) batches, seven BLG (crude herbs) batches, six BLGG batches, and four NBLG batches were collected from different provinces or manufacturers in China, and were numbered from S-01 to S-25 (Table 4). The BLG and NBLG samples were identified by Dr. Li-Hong Wu, and the voucher specimens were deposited in the Herbarium of Shanghai University of Traditional Chinese Medicine (Shanghai, China).

R,S-goitrin was obtained from Shanghai Research and Development Center for Standardization of Traditional Chinese Medicine (Shanghai, China). Hypoxanthine, uridine, guanosine, and adenosine were purchased from the National Institute for Food and Drug control (Beijing, China). Progoitrin, epiprogoitrin, and gluconapin were isolated from the root of *Isatis indigotica* Fort. and their structures were elucidated by mass spectrometry, ^{1}H nuclear magnetic resonance spectrometry, and ^{13}C nuclear magnetic resonance spectroscopy, and were confirmed by comparing the data with those of previous studies. The compound purity was over 98% based on the HPLC area normalization method. The standard structures are illustrated in Figure 6.

HPLC-grade acetonitrile, triethylamine, and formic acid (Tedia, Fairfield, OH, USA) were used for mobile phase preparation. Purified water was prepared with a Mili-Q water-purification system (Millipore, Bedford, MA, USA). Other solvents used for analyses were of analytical grade.

3.2. Instrumentation and Chromatographic Condition

UPLC analysis was performed on a Waters ACQUITY UPLC H-Class system (Waters, Milford, MA, USA) equipped with a binary solvent delivery pump, an auto sampler, and a photodiode array detector (PDA) and was controlled by the Empower-II software. UPLC fingerprinting analysis was carried out at 25 °C on a Waters ACQUITY UPLC BEH C_{18} (1.7 μm, 2.1 mm × 100 mm). A binary gradient elution system composed of acetonitrile (phase A) and 0.1% triethylamine in water (phase B, the pH was adjusted to 4.0 with formic acid) was applied to the fingerprint analysis with the gradient elution as follows: 0 min to 5 min, 0% A; 5 min to 7 min, 0% to 3% A; 7 min to 10 min, 3% A; 10 min to 20 min, 3% to 10% A. The wavelength was set at 254 nm, the mobile flow rate at 0.3 mL min^{-1}, and the on-line UV spectra was recorded in the range of 190 nm to 500 nm.

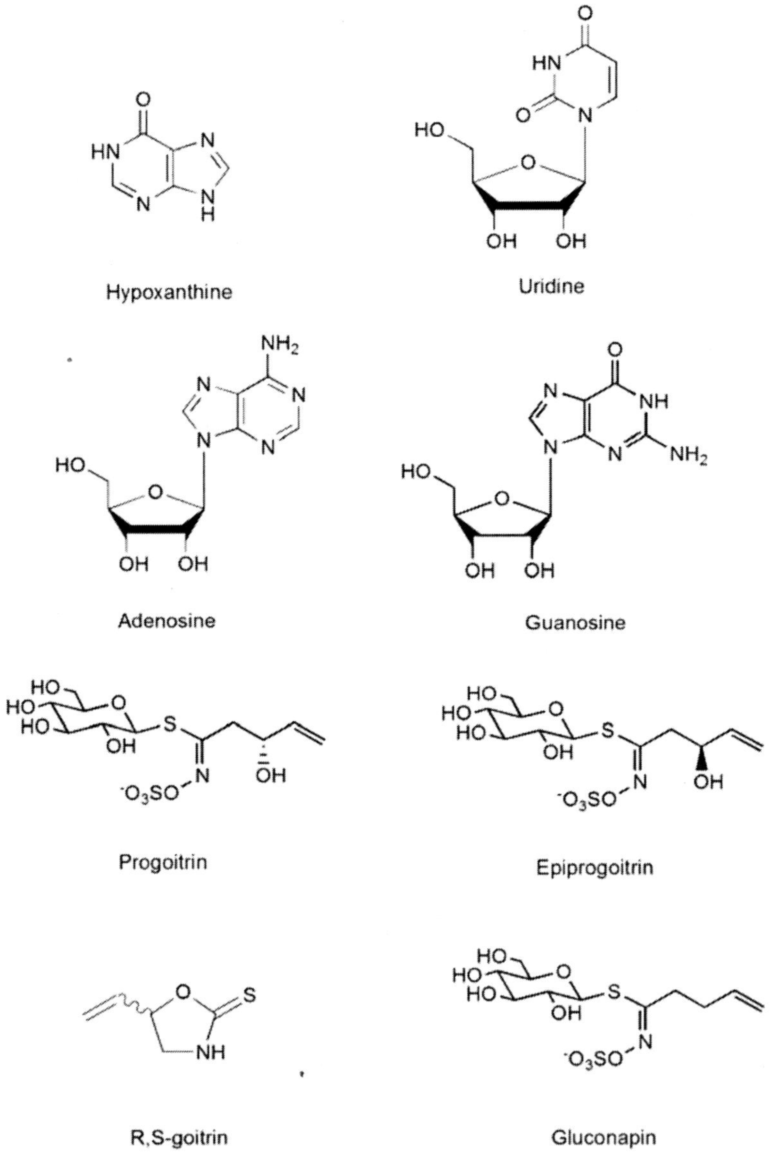

Figure 6. The structures of eight bioactive components in *Radix Isatidis.*

3.3. Preparation of Standard Solution

R,S-goitrin, progoitrin, epiprogoitrin, gluconapin, adenosine, uridine, guanosine, and hypoxanthine CRS were accurately weighed, and dissolved in methanol to produce a solution containing 1.0 mg mL^{-1}, which was used as the reference solution.

3.4. Preparation of Sample Solution

The crude herbs and the prepared BLG and NBLG slices were oven-dried at 50 °C until the weight remained constant. Each sample (1.0 g powder) was extracted with 20 mL of water, which was subjected to reflux for 30 min. The final solution was filtered through a 0.22 μm membrane prior to use. An aliquot of 2.0 μL of each sample solution was injected into the UPLC system for analysis.

3.5. Method Validation

The method was validated for linearity, LOD, LOQ, precision, reproducibility, stability, and accuracy following the International Conference on Harmonization (ICH) guideline and several reports on determination analysis [36,38].

3.6. Data Analysis

Data analysis was performed by the professional software of the Computer-Aided Similarity Evaluation System for Chromatographic Fingerprint of TCM (China Committee of Pharmacopeia, 2004A), which was recommended by the SFDA. The software was employed for synchronization and quantitative comparison between different samples [37].

In addition, the RRT and RPA of each characteristic peak related to the reference peak (*R,S*-goitrin) were also statistically analyzed using the SPSS software package (version 15.0; IBM: Chicago, IL, USA, 2006) and Soft Independent Modeling of Class Analogy (SIMCA)-P (version 13.0; Umetrics: Basel, Switzerland, 2012) [39–41].

4. CONCLUSIONS

In conclusion, the UPLC fingerprint and quantitative analysis of the water-soluble extract of *Radix Isatidis* were first established in this study to determine the eight bioactive constituents in *Radix Isatidis* and its finished products.

Meanwhile, the results from the three chemometric techniques (SA, HCA, and PCA) showed good consistency with one another. The samples from different locations of the same type (crude herbs or prepared slices) were still clustered into one group, and the eight bioactive components were mostly close, which indicates the similarity of internal quality of the samples in the same type. Furthermore, eight marker constituents were found to be specific variables, which could provide the most discrimination and quality control of *Radix Isatidis* and its finished products by quantitative analysis. The PCA loading plot identified the greatest impact factor of characteristic constituents to the classification. In addition, this chromatographic method was efficient and rapid in distinguishing *Radix Isatidis* from other species such as NBLG.

Based on this study, the UPLC fingerprint and quantitative analysis methods provided reliable assurance on systematic and complete quality control on *Radix Isatidis* and its finished products, which would be helpful in improving reasonable developments in *Radix Isatidis* and its finished products.

ACKNOWLEDGMENTS

This research project was supported by the grant from the National Natural Science Foundation of China (No. 81173518, 81172951), the Innovation Program of Shanghai Municipal Education Commission (11YZ70), and the SHUTCM Program for the Establishment of TCM Chemistry.

REFERENCES

1. Pharmacopoeia of People's Republic of China, *The State Pharmacopoeia Committee of People's Republic of China*; Chemical Industry Press: Beijing, China, 2010; Volume 1, pp. 20–21.

2. State Administration of Traditional Chinese Medicine of the People's Republic of China, *Chinese Materia Medica (No.21)*; Shanghai Scientific & Technical Publishers: Shanghai, China, 1999.

3. Wu, P.Y.; Wang, J.M. Botanical study on "Da Qing Ye", "Ban Lan Gen" and "Qing Dai". *Acta Univer. Tradit. Med. Sin. Pharmacol. Shanghai* **1996**, *10*, 50–52.

4. Wu, Q.Z.; Wang, Y.L. Comparative analysis of chemical constituents in north-banlangen and south-banlangen. *Strait Pharm. J* **2007**, *19*, 82–83.

5. Qin, H.Z.; Shi, B.; Li, S.Y.; Fan, X.J.; Song, D.Z.; Xian, H.M. Comparing research in Antiviral action of Rhizoma et *Radix Baphicacanthis Cusiae* and *Radix Isatidis* against a virus. *Chin. Arch. Tradit. Chin. Med* **2009**, *27*, 168–169.

6. Xu, Y.S.; Sun, J.; He, S.Q. Effect of three kinds of *Radix isatidis* preparation on the expression of nucleoprotein of influenza virus. *Shandong Med. J* **2010**, *50*, 8.

7. Liu, S.; Chen, W.S.; Qiao, C.Z.; Zheng, S.Q.; Zeng, M.; Zhang, H.M. Antiviral action of *Radix Isatidis* and *Folium Isatidis*from different germplasm against influenza A virus. *Acad. J. Sec. Mil. Med. Univ* **2000**, *21*, 204–206.

8. Lin, C.W.; Tsai, F.J.; Tsai, C.H.; Lai, C.C.; Wan, L.; Ho, T.Y.; Hsieh, C.C.; Chao, P.D. Anti-SARS coronavirus 3C-like protease effects of *Isatis indigotica* root and plant-derived phenolic compounds. *Antiviral Res* **2005**, *68*, 36–42.

9. Sun, H.H. *Establishment of a Mouse Model for Pandemic H1N1 Influenza Virus and Study on Effect of Banlangen Granules on Mice Challenged with Pandemic H1N1 Influenza Virus*; Peaking Union Medical College: Beijing, China, 2010; pp. 39–49.

10. Wang, Y.T.; Yang, Z.F.; Zhao, H.S.; Qin, S.; Guan, W.D. Screening of anti-H1N1 active constituents from *Radix Isatidis*. *J. Guangzhou Uni. Tradit. Chin. Med* **2011**, *28*, 419–422.

11. Sun, H.H.; Deng, W.; Zhan, L.J.; Xu, L.L.; Li, F.D.; Nv, Q.; Zhu, H. Effect of Banlangen Granules on mice challenged with a/California/ 7/2009. *Chin. J. Comp. Med* **2010**, *20*, 53–56.

12. Cheng, Y.; Li, X.; Chen, J.W.; He, L.W.; Wang, Y.Y. A study on antiviral effects of the active parts from *Radix Isatidis*. *J. Nanjing TCM Univ* **2011**, *27*, 155–157.

13. Ye, W.Y.; Li, Y.; Guo, J.W. Screening of 11 antiviral effect constituents from *Radix Isatidis*. *J. Emerg. Tradit. Chin. Med***2011**, *20*, 1172–1774.

14. Li, X.M. Remarkable progresses in studies on antiviral mechanism researches of *Radix Isatidis* (Baiyunshan, Guangdong). *China News of TCM* **2011**, 8.

15. Liu, S.Y. Tucking up veil the Antiviral causes of *Radix Isatidis*. *China News of TCM* **2011**, A13.

16. Liu, Y.H.; Wu, X.Y.; Fang, J.G.; Xie, W. Chemical constituents from *Radix Isatidis*. *Cent. South Pharm* **2003**, *5*, 302–305.

17. Fang, J.G.; Wang, S.B.; Xu, H.; Liu, Y.H.; Liu, Y.W. Chemical constituents from *Radix Isatidis*. *Chin. Tradit. Herb. Drugs***2004**, *35*, 845–846.

18. He, L.W.; Li, X.; Chen, J.W.; Sun, D.D.; Ju, W.Z.; Wang, C.H. Chemical constituents from water extract of *Radix isatidis.Acta Pharm. Sin* **2006**, *41*, 1193–1196.

19. Wang, Y.Y.; Li, X.; Chen, J.W.; Chen, Y.; Wang, S. Study on Chemical constituents from water extract of *Radix isatidis.Res. Pract. Chin. Med* **2009**, *23*, 54–56.

20. Lee, K.C.; Cheuk, M.W.; Chan, W. Determination of glucosinolates in traditional Chinese herbs by high-performance liquid chromatography and electrospray ionization mass spectrometry. *Anal. Bioanal. Chem* **2006**, *386*, 2225–2232.

21. Xu, L.H.; Huang, F.; Chen, T.; Wu, J. Antivirus constituents of radix of *Isatidis indigotica*. *Chin. J. Nat. Med* **2005**, *3*, 359–360.

22. Deng, X.Y.; Gao, G.H.; Zheng, S.N.; Li, F.M. Qualitative and quantitative of flavonoids in the leaves of *Isatis indigatica*Fort. by ultra-performance liquid chromatography with PDA and electrospray ionization tandem mass spectrometry detection. *J. Pharm. Biomed. Anal* **2008**, *48*, 562–567.

23. Ye, W.T.; Ou-Yang, H.F.; Li, X.L. Study on quality standard of CO Banlangen Granules. *Clin. Med. Eng* **2010**, *4*, 51–52.

24. Kong, W.J.; Zhao, Y.L.; Shan, L.M.; Xiao, X.H.; Guo, W.Y. Investigation on the spectrum-effect relationships of ETOAc extract from *Radix isatidis* based on HPLC fingerprints and microcalorimetry. *J. Chromatogr. B* **2008**, *1*, 109–144.

25. Kong, W.J.; Zhao, Y.L.; Shan, L.M.; Xiao, X.H.; Guo, W.Y. Thermochemical studies on the quantity-antibacterial effect relationship of four organic Accids from *Radix Isatidis* on *Escherichia coli* growth. *Biol. Pharm. Bull* **2008**, *31*, 1301–1305.

26. Kong, W.J.; Zhao, Y.L.; Shan, L.M.; Xiao, X.H.; Guo, W.Y.; Wang, J.B. Determination of salicylic acid in total acid extraction of *Radix Isatidis* by HPLC. *Cent. South Pharm* **2008**, *2*, 137–140.

27. Fan, L.F.; Zhang, L.T.; Yuan, Z.F.; Xu, H.J.; He, W. HPLC determination the contents of indigo and indirubin in *Radix Isatidis. Chin. J. Pharm. Anal* **2008**, *4*, 540–543.

28. Na, G.X.; Fang, P. Research overview in degradation of glucosinolates. *Food Sci* **2008**, *29*, 350–354.

29. Hashem, F.A.; Saleh, M.M. Antimicrobial components of some cruciferae plants (*Diplotaxis harra* Forsk. and *Erucaria microcarpa Boiss.*). *Phytother. Res* **1999**, *13*, 329–332.

30. Wang, R.; Yang, H.Y.; Yang, Q.W. Study on quality standard of *Isatidis Radix. Chin. Tradit. Herb. Drug* **2010**, *41*, 478–480.

31. Xie, Z.Y.; Shi, Y.H.; Wang, Z.T.; Wang, R.; Li, Y.M. Biotransformation of glucosinolates epiprogoitrin and progoitrin to (*R*)- and (*S*)-goitrin in *Radix isatidis*. *J. Agric. Food. Chem* **2011**, *59*, 12467–12472.

32. Zhou, W.; Xie, M.F.; Zhang, X.Y.; Liu, T.T.; Duan, G.L. Improved liquid chromatography fingerprint of fat-soluble*Radix Isatidis* extract

using multi-wavelength combination technique. *J. Sep. Sci* **2011**, *34*, 1123–1132.

33. Xiao, X.; Luo, G.A.; Wang, Y.M. Fingerprints of *Isatidis Radix* by HPLC. *J. Jiangxi Univ. TCM* **2010**, *22*, 67–69.

34. Zou, P.; Hong, Y.; Koh, H.L. Chemical fingerprinting of *Isatis Indigatica* root by RP-HPLC and Hierarchical clustering analysis. *J. Pharm. Biomed. Anal* **2005**, *38*, 514–520.

35. Shi, Y.H.; Xie, Z.Y.; Wu, Y.C.; Li, Y.M.; Wang, R.; Wang, Z.T. Determination of *R,S*-goitrin of Banlangen preparations by RP-HPLC. *Chin. J. Exp. Tradit. Med. Form* **2010**, *17*, 144–146.

36. Tang, D.Q.; Yang, D.Z.; Tang, A.B.; Gao, Y.Y.; Jiang, X.G.; Mou, J.; Yin, X.X. Simultaneous chemical fingerprint and quantitative analysis of *Ginkgo biloba* extract by HPLC-DAD. *Anal. Bioanal. Chem* **2010**, *396*, 3087–3095.

37. Wei, H.; Sun, L.N.; Tai, Z.G.; Gao, S.H.; Xu, W.; Chen, W.S. A simple and sensitive HPLC method for the simultaneous determination of 8 bioactive components and fingerprint analysis of *Schisandra sphenanthera*. *Anal. Chim. Acta* **2010**,*662*, 97–104.

38. Jin, X.F.; Lu, Y.H.; Wei, D.Z.; Wang, Z.T. Chemical fingerprint and quantitative analysis of *Salvia plebeian* R.Br. by high-performance liquid chromatography. *J. Pharm. Biomed. Anal* **2008**, *48*, 100–104.

39. Li, Y.; Wu, T.; Zhu, J.J.; Wan, L.L.; Yu, Q.; Li, X.X.; Cheng, Z.Z.; Guo, C. Combinative method using HPLC fingerprint and quantitative analyses for quality consistency evaluation of an herbal medicinal preparation produced by different manufactures. *J. Pharm. Biomed. Anal* **2010**, *52*, 597–602.

40. Wang, Y.X.; Li, Q.; Wang, Q.; Li, Y.J.; Ling, J.H.; Liu, L.L.; Chen, X.H.; Bi, K.S. Simultaneous determination of seven bioactive components in Oolong Tea *Camellia sinensis*: Quality control by chemical composition and HPLC fingerprints. *J. Agric. Food Chem* **2012**, *60*, 256–260.

41. Wu, Q.Y.; Zhou, Y.; Jin, X.; Guan, Y.; Xu, M.; Liu, L.F. Chromatographic fingerprint and the simultaneous determination of five bioactive components of *Geranium carolinianum* L. water extract by high performance liquid chromatography. *Int. J. Mol. Sci* **2011**, *12*, 8740–8749.

8

AL(GA)INP-GAAS PHOTODIODES TAILORED FOR SPECIFIC WAVELENGTH RANGE

Yong-gang Zhang[1] and Yi Gu[1]

[1] State Key Laboratory of Functional Materials for Informatics, Shanghai Institute of Microsystem and Information Technology, Chinese Academy of Sciences. Shanghai, China

1. INTRODUCTION

The sun light shining on our earth, with its main energy concentrated in visible extending to ultraviolet (UV), has activated this planet adequately. Therefore, the spectral response features of life-forms including human beings, as well as plentiful artificial creatures, are linked to those bands spontaneously and tightly. Photodiodes or photodetectors (PDs) with specific wavelength response in UV-visible bands have many practical and potential applications including ocean or water related sensing and communication, medical engineering and photodosimetry, missile guidance and countermeasures, and so on. In those wavelength bands, various groups of II-VI, III-V and VI materials could be utilized for PDs. Among them, Si PD should be the most successful one, whereas its wide response extends to near-infrared inherently, also, the performance of Si PD is limited by its indirect and relatively narrow bandgap, which restricts its applications in certain cases. Among various optional materials in those bands, the III-V Al(Ga)InP system, especially the ternary AlInP and GaInP, may work well in visible or even extending to UV band. $Al_{0.52}In_{0.48}P$ and $Ga_{0.51}In_{0.49}P$, which are lattice matched to GaAs substrate, have band gaps about 2.3 eV and 1.9 eV respectively, the combination of those two ternary materials, in conjunction with the quaternary AlGaInP system, also gives a big room in tailoring the response of the photodiodes to a specific wavelength region. Besides, profiting from the wider bandgap comparing to that of Si, higher working temperature, lower dark current and better radiation hardness could be expected for those robust materials. Furthermore, for this GaAs based III-V non-nitride system, the doping in both n and p type is feasible, a quite mature growth and processing technology can be relied on, so photovoltaic detectors and arrays with better performance could be presumed.

In this chapter, a simple review on the material issues of PDs in these bands will appear first, then concentrated on the gas source molecular beam epitaxy (GSMBE) growth of the ternaries lattice matched to GaAs substrates, including GaInP and less-studied AlInP. The doping, structural and optical properties of the GSMBE grown materials will be investigated in detail, then turning into the growth of PD structures and processing of the device chips composing of those ternaries, mainly based on our experience. Finally, the performance of developed AlInP-GaInP-GaAs and AlInP-GaAs photodiodes tailored for specific wavelength range will be characterized and discussed.

2. MATERIAL ISSUES

2.1. Uv-Visible Detector Materials Overview

In UV-visible bands normally quantum or photon type PDs, instead of thermal type PDs, are preferable. The materials for quantum type PDs sensitive to UV-visible bands could be chosen from group IV (e.g.: Si, SiC), group III-V (nitride, phosphide or arsenide), group II-VI (e.g.: CdS, CdTe) and so on, mainly depending on their bandgap (e.g. Razeghi et al. 1996). The expected sensitive wavelength region of a PD is mainly depending on its application requirements. Normally, inside the anticipant sensitive wavelength region a high response of the detector is expected. However, outside the anticipant sensitive wavelength region a full cutoff of the PD is preferable, because in this case the response is not a signal but noise or interference to the applications. For a quantum type PD operating in certain wavelength region, regardless of artificial resonate or filtering structures, the cutoff wavelength at long wavelength side is determined by the bandgap of the sensitive material inherently. The response of the detector at short wavelength side is depending on the material properties, device structures and so on. Normally, it will not cutoff sharply but drop continuously until reaching an unacceptable level. Furthermore, a better performance of the detector could be expected for a material with larger bandgap, regardless of some other effects. Normally the introduction of artificial resonate or filtering structures on the detection system, regardless of internally or externally, will increase the complexity, as well as cause losses, therefore a PD just matched to desired wavelength region inherently is strongly expected. Based on above considerations, when the cutoff wavelength of the detector at long wavelength side is fixed, as a rule of thumb a material with close to but slightly longer bandgap wavelength is expected. After that, the response of the detector could be finely tailored and optimized using different material stacks, device structures and processing parameters.

Quantum type PDs may have different types such as photoconductive or photovoltaic; from practical point of view photovoltaic types are more preferable because the detectors could be operated without bias, and therefore with lower dark current. In visible wavelength range group IV Si is definitely the most successful material to demonstrate photovoltaic detectors. However, its response is wider and extends to both near infrared (NIR) and UV sides. Therefore for certain applications in specific wavelength region a filter or resonate structure has to be applied to suppress the response in both long and short wavelength sides. Also, the indirect and relative narrow bandgap will limit its performance. The CdS detector has no infrared response inherently, whereas currently it only works as a photoconductor and seems hard to reach high performance and high speed. Some other wide bandgap II-VI, SiC and III-V nitrides have also attracted much attention in those bands (e. g. Ando et al. 2002;Zhang et al. 1997; Zhang et al. 2009; Rigutti et al. 2010). Especially, various nitride AlGaN/GaN detector structures including positive-intrinsic-negative (PIN) (e. g. Pulfrey et al. 2001), Schottky (e. g.Lee et al. 2012), metal-semiconductor-metal (MSM) (e. g. Mosca et al. 2004) or metal-insulator-semiconductor (MIS) (e. g. Chang et al. 2006) have been adopted. However, their responses are mainly restricted in the UV side, and the tailoring of the response width is difficult. In addition to above-mentioned material systems, the III-V AlGaInP material system, especially the ternary AlInP and GaInP, may also work well in visible and may even extend to UV band (e. g. Zhang et al. 2005,2010; Li et al. 2011).

The calculated lattice constant and bandgap energy contours of the $Al_zGa_xIn_yP$ (x+y+z=1) quaternary system are shown in Figure 1. In this figure the sloping solid line and almost aclinic dot line indicate the bandgap energies and lattice constants of this quaternary respectively. In the calculation the energy bowing of the ternary was taken into account, whereas the bowing of the quaternary was ignored. This quaternary system is composed of three group-III elements of Al, Ga and In, with only one group-V element of P. In this material system the ternaries $Al_zIn_{1-z}P$ and $Ga_xIn_{1-x}P$ are at the left and right ridges; the ternary $Al_zIn_{1-z}P$ could be recognized as the alloy of binaries indirect AlP and direct InP, and ternary $Ga_xIn_{1-x}P$ as the alloy of binaries GaP and InP also with indirect and direct bandgap, respectively. The direct bandgap of $Al_zIn_{1-z}P$ and $Ga_xIn_{1-x}P$ at 300 K could be written as $E_g=3.55z+1.34(1-z)+0.48z(1-z)$ and $E_g=2.78x+1.34(1-x)-0.65x(1-x)$ respectively (Vurgaftman et al. 2001). Those ternaries have the widest direct bandgap among common III-Vs excepting nitrides.

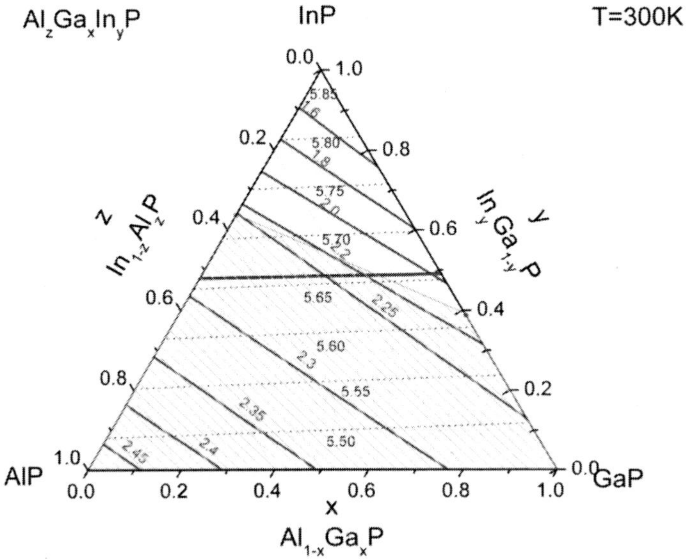

Figure 1. Lattice constant and bandgap energy of $Al_zGa_xIn_yP$ quaternary system. The shadowy bottom region indicates the indirect bandgap zone; the thick solid line in the middle indicates the quaternary alloy lattice matched to GaAs substrate.

Figure 2 shows the bandgap energy and band edge wavelength of the AlInP and GaInP ternary alloys versus the indium composition x at 300 K. The materials with direct and indirect bandgap are shown by solid and dashed lines, respectively. The indium composition of the ternaries latticed matched to GaAs has also been indicated. The lattice constants of AlP and GaP are very close, which makes the indium compositions for

GaAs-lattice-matched AlInP and GaInP are also very close. $Al_xIn_{1-x}P$ is lattice matched to GaAs substrate when the indium composition is around 0.48, while ternary $Ga_xIn_{1-x}P$ is lattice matched to GaAs substrate with indium composition around 0.49. It makes ternary $Al_{0.52}In_{0.48}P$ and $Ga_{0.51}In_{0.49}P$ important epitaxial materials from both research and application points of view. The $Al_{0.52}In_{0.48}P$ on GaAs shows many similar characteristics as the AlGaAs/GaAs or AlInAs/InP system. It has an indirect bandgap about 2.3 eV, which is the largest bandgap among the practical non-nitride III-V alloys, so a great deal of applications could be expected on this wide bandgap material (Gu et al. 2006, 2007). The bandgap of 2.3 eV corresponds to a cutoff wavelength of about 540 nm, which is located in the visible band. The $Ga_{0.51}In_{0.49}P$ on GaAs has a direct band gap of 1.9 eV, corresponding to the wavelength of about 650 nm. The combination of AlInP and GaInP two materials gives a room in tailoring the response of the photodiodes to a specific wavelength region in visible band and even extending to UV, but blind to infrared inherently.

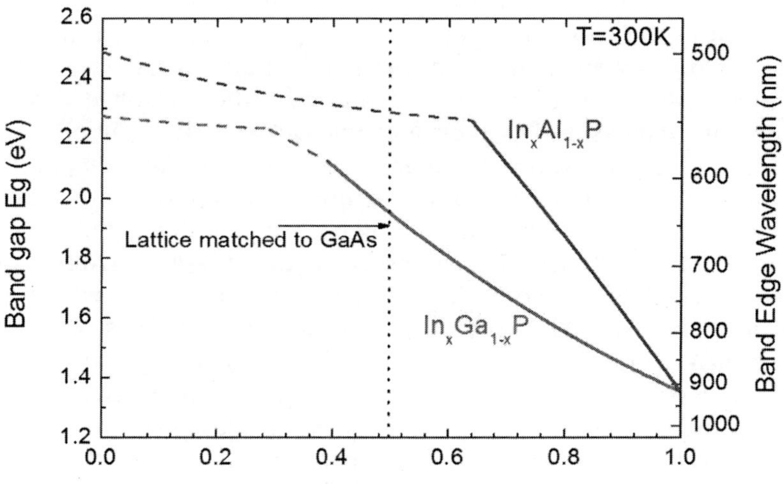

Figure 2. Bandgap energy and related band edge wavelength of $In_xGa_{1-x}P$ and $In_xAl_{1-x}P$ ternary alloys versus the indium composition x at 300 K. The materials with direct and indirect bandgap are shown by solid and dotted lines, respectively. The indium composition when the ternaries latticed matched to GaAs has also been indicated.

2.2. Gas Source Mbe Growth Of Gainp and Alinp Ternary Alloys

In our growth process, a VG Semicon V80H GSMBE system was applied. The best background vacuum achieved in this system was about 1×10^{-11} Torr. The elemental indium and gallium Thermacells, as well as aluminum standard effusion cell, were used as group III sources, and their fluxes were controlled by changing the cell temperatures. Arsine (AsH_3) and phosphine (PH_3) high-pressure cracking cells were used as group V sources, their fluxes were pressure controlled, and the cracking temperature was around 1000 °C. Standard beryllium and silicon effusion cells were used as p- and n- doping sources, and the doping level was also controlled by changing the cell temperatures. Before the growth, the fluxes of group III sources were calibrated by using an in situ ion gauge.

The GaInP and AlInP layers were all grown on (100) oriented GaAs semi-insulating (S. I.) epi-ready substrates, which were placed on In-free Mo blocks. Prior to the growth, the surface oxide desorption of the substrate was carried out under As_2 flux. This process involved a slow ramp-up of the substrate temperature until the reflection high energy electron diffraction (RHEED) pattern shown an abrupt transformation to 2×4 surface reconstruction, in this case the substrate desorption temperature measured by thermocouple was usually about 630 °C, then the substrate temperature was decreased to appropriate temperature to begin the growth. Prior to the growth, an appropriate As-P flux exchange procedure with pump down of arsenic for 1 minute before open P flux was used. The pressure in the growth chamber during growth, which is related to the flux of group V sources, was adjusted to about 2×10^{-5} Torr. The growth rate of the epi-layers was controlled to be about 1 μm/hour, and the epi-layer thickness of all grown samples in this study was around 1 μm. After growth, the morphology of the grown samples was observed using a Normasky microscope, their structural characteristics were measured by using a Philips X-pert type X-ray diffractometer (XRD) including a Ge (220) four-folded monochrometer, and the carrier concentration of the epi-layer was determined by using Hall measurement in Van der Pauw scheme or electrochemical capacitance-voltage (EC-V) profiler at room temperature.

Ternary GaInP material has been widely applied in visible laser diodes (e.g.: Kaspari et al. 2008) and solar cells (e.g.: Geisz et al. 2008) structures. Many efforts have been paid to the epitaxy of GaInP, and found that the ordered degree plays a critical role in the growth. In our GSMBE growth, the calibration growth of GaInP layers were performed as the first step, the indium and gallium source temperatures were adjusted to grow $Ga_{0.51}In_{0.49}P$, which was lattice matched to GaAs substrate. The lattice mismatch between epilayers and substrate were confirmed by XRD measurements, the typical XRD rocking curve of grown $Ga_{0.51}In_{0.49}P$ on GaAs is shown in Figure 3. Since the thermal expansion coefficient of $Ga_{0.51}In_{0.49}P$ is larger than that of GaAs, the amount of lattice mismatch determined at room temperature is shifted to positive side comparing to the lattice match at growth temperature; therefore, at room temperature a slightly positive lattice mismatch should be preferable, the estimated lattice mismatch of $+5\times10^{-4}$ at room temperature may lead to the precise lattice match of $Ga_{0.51}In_{0.49}P$ to GaAs at growth temperature. As shown in Figure 3, the XRD measurement shows that this optimized GaInP epi-layer has a positive mismatch around $+4.9\times10^{-4}$ to GaAs substrate, with a FWHM of 28.6 arcsec for the epi-layer and 18.6 arcsec for the substrate.

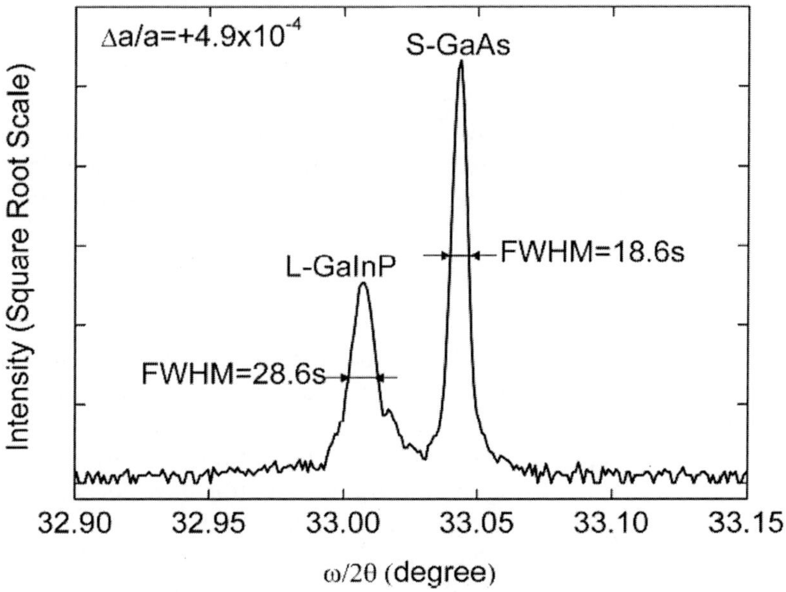

Figure 3. Measured X-ray rocking curve of typical GSMBE grown $Ga_{0.51}In_{0.49}P$ on (100) GaAs substrate.

After that, a series of $Ga_{0.51}In_{0.49}P$ layers were grown with different silicon doping levels, including an unintentionally doped sample and three samples with silicon cell temperatures of 1130 °C, 1150 °C and 1200 °C. The room temperature electron concentrations measured by Hall effects were high resistance, 2.4×10^{17} cm^{-3}, 1.1×10^{18} cm^{-3} and 4.6×10^{18} cm^{-3}, respectively. Photoluminescence (PL) measurements were performed for these samples and shown in Figure 4. The samples were mounted in a continuous-flow variable-temperature helium cryostat (using a HC-2D-1 APD Cryogenic System and a Lakeshore 330 temperature controller). A Coherent INNOVA 305 argon-ion laser (=514.5 nm) was used and the laser beam was focused to a spot size of approximately 0.5 mm^2 on the sample to excite the PL spectra. The PL light was collected and collimated by a couple of large quartz lens, and then focused into a Jobin Yvon THR1000 monochrometer with a 1200 grooves/mm grating, the PL signal was detected by using a photo-multiplier-tube (PMT).

It is shown in Figure 4 that the PL emission keeps nearly the same for the samples with unintentional or lower doping, the PL peaks are at 1.951 eV and the full width half-maximum (FWHM) are about 24 meV. However, as the doping level is increased to about 1×10^{18} cm^{-3}, the PL emission wavelength is blue shifted to about 1.963 eV and the peak FWHM increases to 38 meV. When the doping level is further increased to about 5×10^{18} cm^{-3}, the PL wavelength is further blue shifted to about 1.972 eV and the FWHM further increases to 58 meV. Moreover, the PL intensity for the sample with the highest silicon doping level is decreased significantly to about 1/10. The blue shift, broad FWHM and decreased intensity can be explained by the decreased ordered degree, which could be attributed to the high silicon doping (Yoon et al. 1999; Longo et al. 2001).

In the past years, epitaxial AlInP lattice matched to GaAs has been studied for the applications such as window/anti-reflection (AR) layer of GaInP/GaAs multi-junction solar cells (Takamoto et al. 1997;Karam et al. 2001) and cladding layer of visible lasers and modulators (Murata et al. 1991), mainly by using metal-organic chemical vapor deposition (MOCVD) techniques. The GSMBE growth of relative thin AlInP layers for the tunnel junction or window/AR layer of tandem solar cells have also been reported (Li et al. 1998). Recently, AlInP has been found with great potential to demonstrate very low dark current avalanche photodiodes (APD) (Ong et al. 2011). However, comparing with the well-developed GaInP growth, the study of the growth process and doping characteristics of AlInP on GaAs remains quite insufficient, especially for the growth of thicker and composition diversified layers. Compared to MOCVD, MBE yields a higher doping efficiency and less diffusion effects for its lower growth temperatures, and the use of gas source for group V elements in GSMBE makes the process more realizable, especially for the growth of phosphide (Gu et al. 2006& 2007).

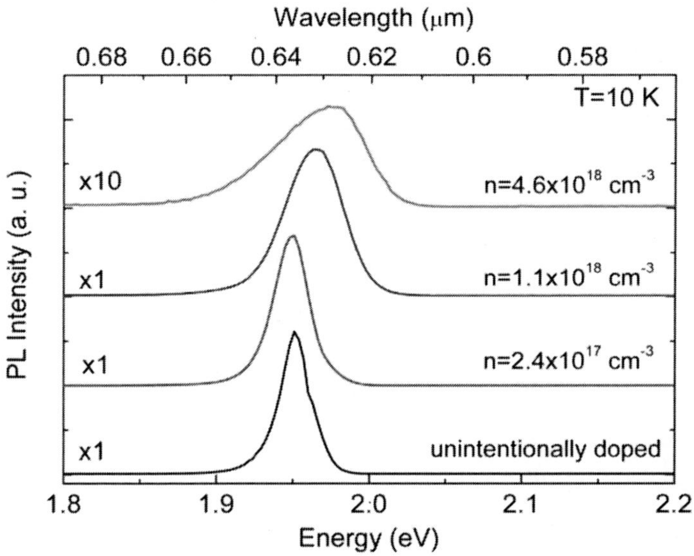

Figure 4. The PL spectra of $Ga_{0.51}In_{0.49}P$ samples with different electron concentration levels at 10 K.

During our GSMBE growth of ternary AlInP layers, the fluxes of aluminum and indium and their ratio in the growth of $Al_xIn_{1-x}P$ were investigated in detail. Figure 5 shows the XRD measured AlP mole fraction of AlInP versus aluminum flux ratio during the growth. The aluminum flux ratio (η_{Al}) is defined as $\eta_{Al}=f_{Al}/(f_{Al}+f_{In})$, in which f_{Al} and f_{In} is the flux of aluminum and indium cell respectively. The AlP mole fraction shows a quite linear function with η_{Al} in our experimental range, which could be used to predict the AlP mole fraction before sample growth. The non-unity slope of the line shows that indium and aluminum elements have quite different sticking coefficient, at this growth temperature the sticking coefficient of aluminum is much higher than that of indium.

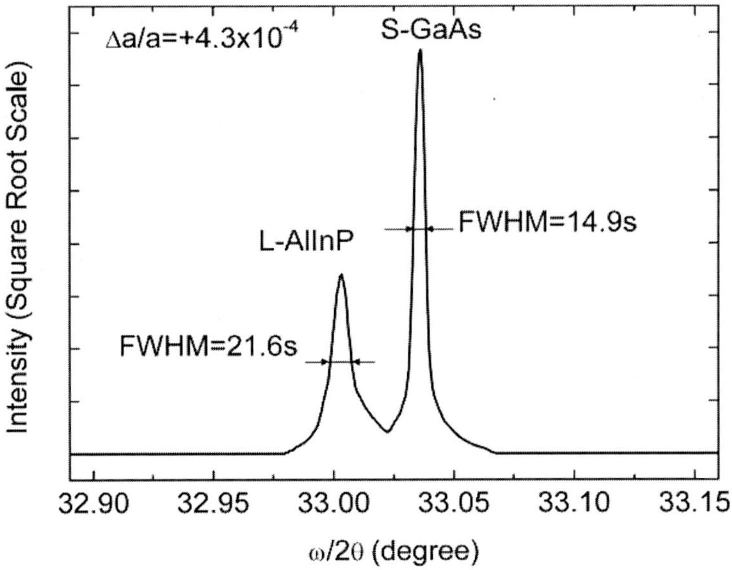

Figure 5. The AlP mole fraction versus the aluminum flux ratio η_{Al} in the GSMBE growth of AlInP.

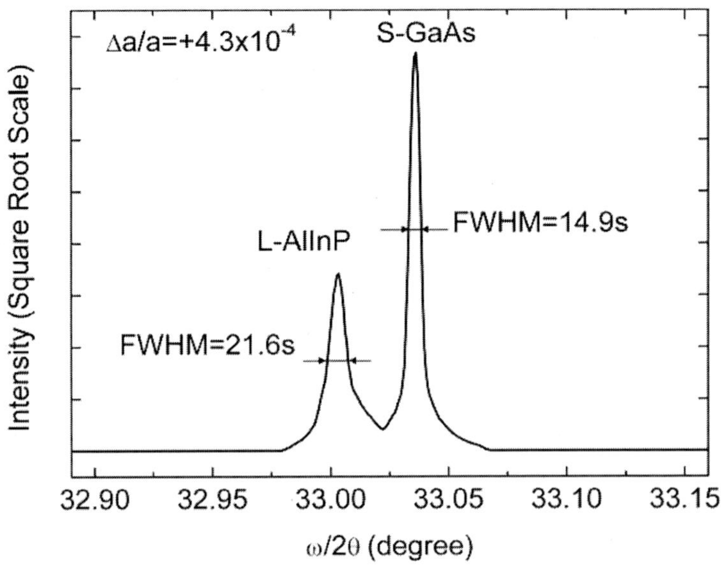

Figure 6. Measured X-ray rocking curve of typical GSMBE grown $Al_{0.52}In_{0.48}P$ on (100) GaAs substrate.

Similar to GaInP/GaAs, a slightly positive lattice mismatch of AlInP epi-layer with respect to GaAs substrate should be preferable at room temperature. The XRD measurement shows that the optimized AlInP epi-layer has a positive mismatch around $+4.3\times10^{-4}$ to the GaAs substrate as shown in Figure 6, with a FWHM of 21.6 arcsec for the epi-layer and 14.9 arcsec for the substrate, which are among the best for epitaxial grown layers.

In this condition, a perfect mirror-like surface could be reached as shown in Figure 7(a). The Normasky micrographs of other grown samples with larger lattice mismatch to GaAs are shown inFigure 7 (b-d). Figure 7(b) shows the micrograph of AlInP epi-layer with large negative mismatch around -4.7×10^{-3}, in this condition light ripple pattern could be seen along $<01\bar{1}>$ direction. At even larger negative mismatch of -6.1×10^{-3}, a cross ripple pattern along both $<011>$ and $<01\bar{1}>$ direction could be seen as shown in Figure 7(c). In negative mismatch conditions, the AlInP epi-layer is affected by tensile strain from the substrate; in this case the misfit dislocation may develop, and leads to poor morphology of the epi-wafer. However, the positive mismatched sample shows a mirror-like surface even if the mismatch reaches as large as $+7.3\times10^{-3}$ as shown in Figure 7(d). It could be deduced that the compressive strain during the growth has much weak effects on the surface morphology.

XRD rocking curves were measured for AlInP epi-layers with different AlP mole fractions, and the FWHM of XRD epi-layer and substrate peaks as a function of AlP mole fraction is shown in Figure 8. The AlInP samples in the condition of around lattice match have narrower FWHM values and show the better lattice quality. The FWHM of epi-layer peaks kept below 100 arcsec as the lattice mismatch was in the range of about -4×10^{-3} to $+3\times10^{-3}$, and was broadened abruptly when out of this range. The FWHM of substrate peaks was broadened to about 160 arcsec as the negative lattice mismatch was increased to -6×10^{-3}, but was still narrower than 40 arcsec as the positive lattice mismatch was increased to $+7\times10^{-3}$. It could be deduced that AlInP epi-layer with positive mismatch has better lattice quality than those with negative mismatch. This behavior also suggests that the compressive strain has weaker effects on lattice quality than the tensile strain during the growth.

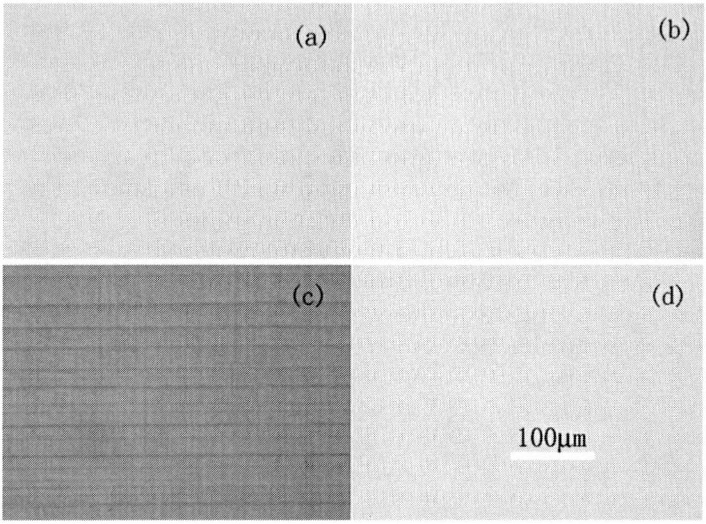

Figure 7. Normasky micrographs of the GSMBE grown AlInP epi-layers with different lattice mismatch to GaAs: (a) nearly lattice-matched layer with positive mismatch of $+4.3\times10^{-4}$; (b) layer with negative mismatch of -4.7×10^{-3}; (c) layer with even large negative mismatch of -6.1×10^{-3}; (d) layer with positive mismatch of $+7.3\times10^{-3}$.

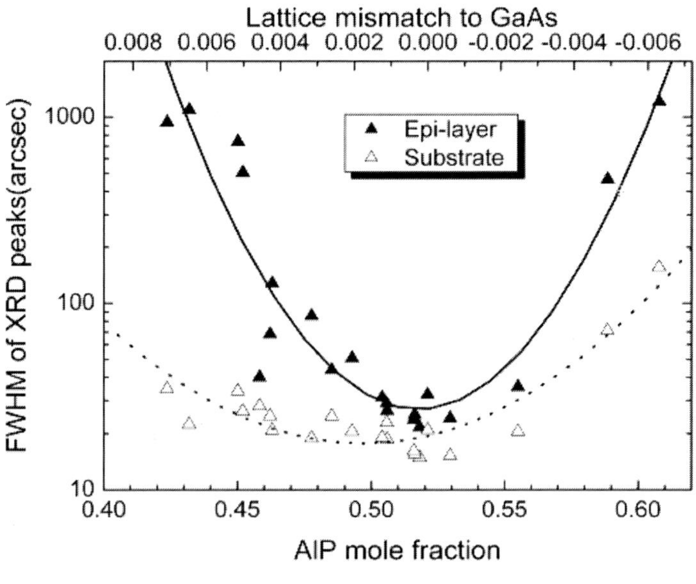

Figure 8. FWHM of XRD peaks for AlInP epi-layers and substrates versus the AlP mole fractions.

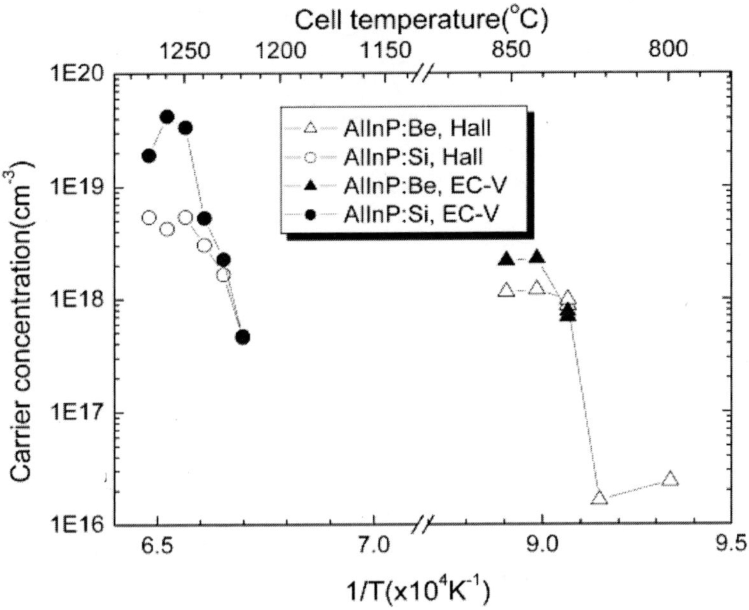

Figure 9. Average (carrier density divided by epi-layer thickness) carrier concentration in AlInP layers versus reciprocal temperature of Si and Be cells.

The electrical properties of AlInP were also investigated. Figure 9 shows the measured electron or hole concentration in Si or Be doped AlInP samples as a function of the reciprocal temperature of Si and Be cells, all those samples were around the lattice matched conditions of x=0.52. Considering that the production of two dimensional electron or hole gas is possible for higher bandgap AlInP layer grown on GaAs substrate with lower bandgap, the EC-V profiling was also done using a PN4300 Profiler, besides the Hall measurements. Regardless the measurement and system error (with a coefficient of about 2) for different measurement techniques, a good correlation between the EC-V and Hall results definitely exists. It was observed that the Hall hole concentration maintained around 10^{16} cm^{-3} as the Be cell temperature (T_{Be}) was below 820 °C, and increased to 1.2×10^{18} cm^{-3} when T_{Be} was increased to 840 °C. Similarly, as the Si cell temperature (T_{Si}) was below 1200 °C the Hall electron concentration was quite lower, whereas as T_{Si} increased to 1250 °C the electron concentration also increased to 5.4×10^{18} cm^{-3} abruptly. At higher cell temperatures a saturation phenomenon was observed.

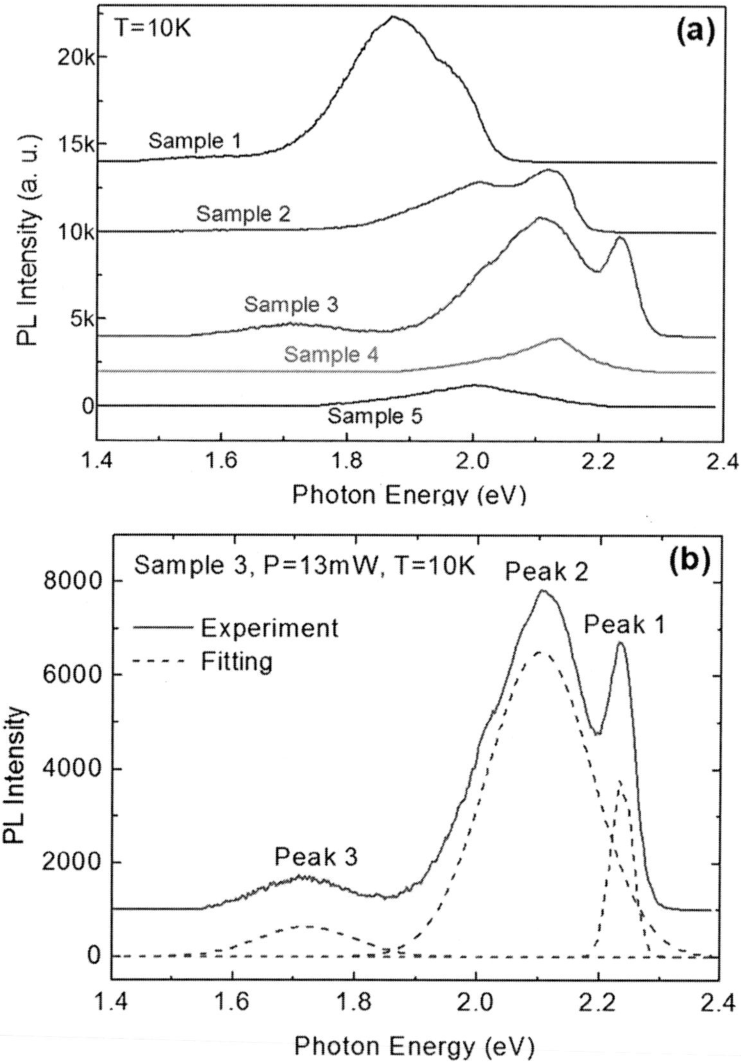

Figure 10. a) The PL spectra of five AlInP samples with different aluminum compositions at 10 K; (b) Three peaks were defined by using Gaussian fitting for PL spectrum of sample 3 at 10 K.

The optical properties of the grown AlInP samples were studied by using PL measurements, and the mechanisms were analyzed in detailed. Figure 10(a) shows the PL spectra of five AlInP samples with different aluminum composition at 10 K, for consistency the same excitation power of 13 mW was used. For sample 1-5, the measured AlP mole fractions by XRD using strained estimation (X_{Al-S}) are 0.38, 0.40, 0.42, 0.47 and 0.51 respectively; whereas using relaxed estimation the AlP mole fractions (X_{Al-R}) are 0.22, 0.28, 0.32, 0.42 and 0.49 respectively. The real AlP mole fractions of these samples should be between the strained and relaxed values. Sample 1, 2 and 3 have direct bandgap, sample 5 has indirect bandgap, and sample 4 is near the critical condition. The lattice-mismatch between AlInP and GaAs substrate of sample 1 is the largest, and decreases in turn for sample 2, 3, 4, 5. As shown inFigure 10(a), several overlapped peaks, which should have different origins, could be seen in the spectra. For sample 1-3, two main peaks appear clearly, in sample 3 a weak peak also exits at lower-energy side. As the two strong peaks overlapped each other, Gaussian fitting was used to determine the peak energy and intensity exactly. In the fitting two peaks were used for sample 1 and 2, whereas three peaks for sample 3. The fitting result for sample 3 is shown in Figure 10(b), the two strong peaks are marked as peak 1 and peak 2, the weak side-peak is marked as peak 3.

To study the origins of these emission peaks, excitation power dependence of the PL spectra was performed on sample 3 from 1 mW to 180 mW at 10 K as shown in Figure 11(a). These spectra were fitted by using three Gaussian peaks, and Figure 11(b) shows the peak energy for the three peaks as a function of the excitation power. It could be seen that peak 2 shifts to higher energy side (blue shift) as the excitation power increases, whereas no obvious shift is observed for the other two peaks. Therefore, peak 2 is considered due to shallow donor-acceptor (D-A) pair recombination, and this PL shift can be explained in terms of the Coulomb interaction of donors and acceptors as a function of their separation. As the excitation power increases, there are more probabilities for the transitions between closer D-A pairs; therefore, the emission intensity increases, and the transition energy tends to move towards higher energy side [Zacks et al. 1972]. On the other hand, peak 1 is deemed to be related to band-to-band (B-B) transition, which shows narrow width and not shifts as the excitation power increases. The side-peak around 1.72 eV may be considered as associated with the presence of arsenic related defects according to its peak energy and weak intensity, which may be imported by the growth background. The PL intensity and FWHM of B-B transition peak in sample 1, 2 and 3 versus the AlP mole fraction is shown in Figure 12. As AlP mole fraction increases, the intensity of B-B transition peak is increased and the FWHM is decreased, which is due to the reduction of nonradiative recombination centers and increase of the radiation efficiency in smaller lattice-mismatched material.

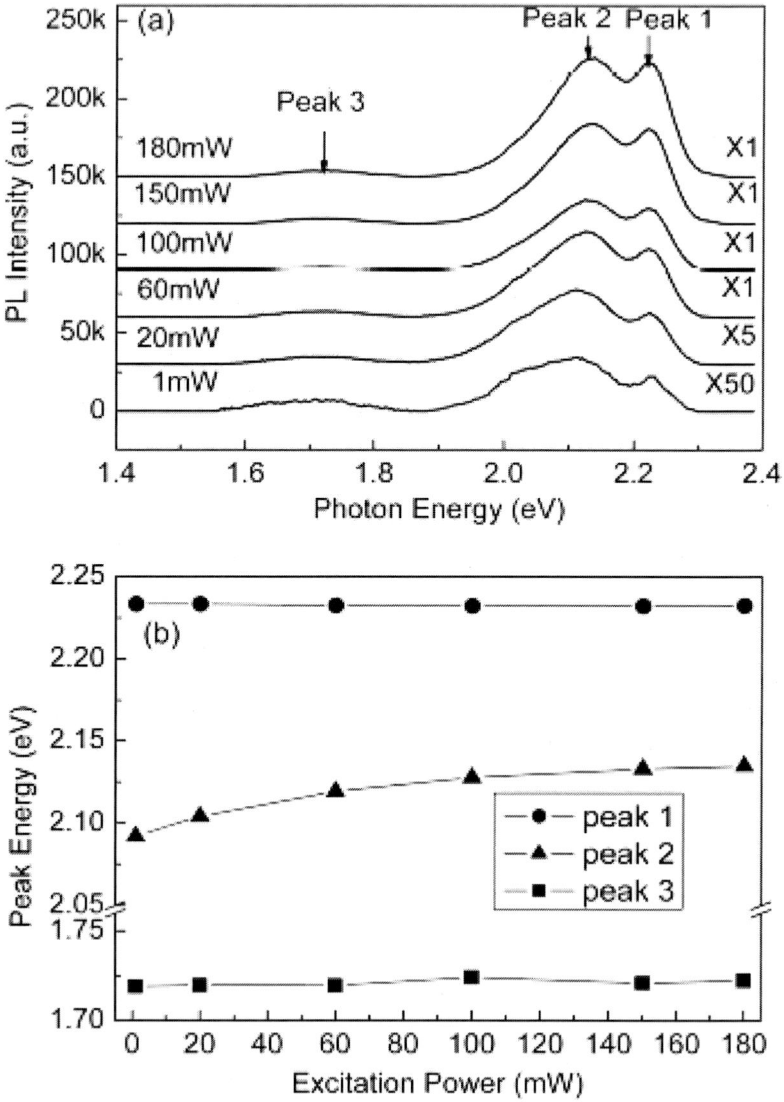

Figure 11. a) Excitation power dependence of the PL spectra for AlInP sample 3 at 10K; (b) Peak energy of the three Gaussian fitting peaks versus the excitation power for AlInP sample 3 at 10 K.

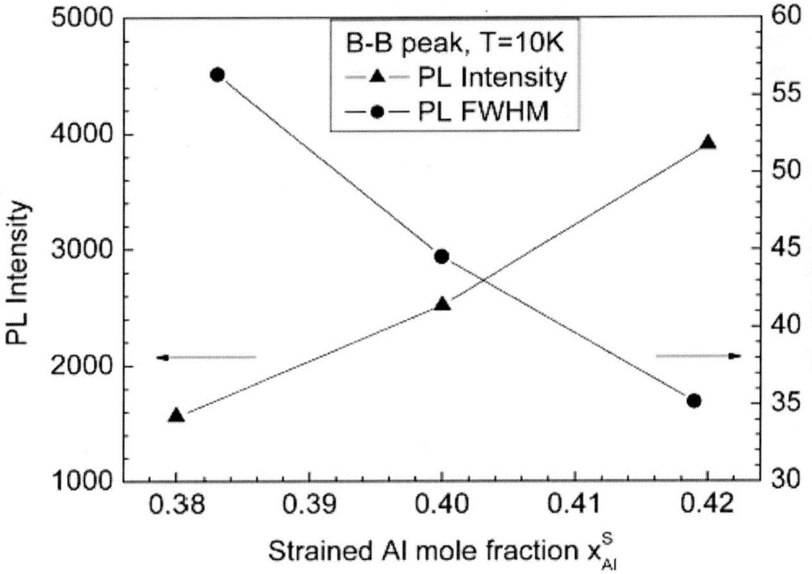

Figure 12. PL intensity and FWHM of B-B peaks for AlInP Sample 1-3 at 10K versus strained AlP mole fraction $X_{Al\text{-}S}$ at excitation power of 13 mW.

Table 1 lists the peak energies of the two strong peaks of B-B related E_{BB} or D-A related E_{DA} in the PL spectra, the separation energy between B-B and D-A transition peaks (E_{BB}-E_{DA}) for sample 1-3 are also listed. For sample 1-3 both the B-B and D-A transition peaks shift to higher energy side (blue shift) as AlP mole fraction increases, which is mainly caused by the increase of the bandgap. By using $X_{Al\text{-}S}$ and $X_{Al\text{-}R}$ respectively, the theoretical bandgap energies at 10 K (written as $E_{G\text{-}S}$ and $E_{G\text{-}R}$ for short) of all samples were calculated and also listed in Table 1, for sample 1-4 the Γ-valley determines the bandgap, whereas for sample 5 the X-valley has the lowest energy. For sample 1-3, the rough fraction of strained portion in the epi-layers (η) could be calculated by the linear interpolation using the following equation:

$$E_{G-S} \cdot \eta + E_{G-R}(1-\eta) = E_{BB} \tag{1}$$

The calculated results of are also listed in Table 1. Results show that this fraction increases as expected when the Al mole fraction increases and lattice mismatch decreases.

Table 1. Measured and calculated data of grown AlInP samples.

Sample No.	X_{Al-S}	X_{Al-R}	E_{BB} (eV)	E_{DA} (eV)	E_{BB}-E_{DA} (meV)	E_{G-S} (eV)	E_{G-R} (eV)	η (%)
1	0.38	0.22	1.98	1.87	104	2.15 (Γ)	1.83 (Γ)	47
2	0.40	0.28	2.12	2.01	117	2.19 (Γ)	1.95 (Γ)	71
3	0.42	0.32	2.23	2.10	140	2.24 (Γ)	2.02 (Γ)	95
4	0.47	0.42	NA	2.14	NA	2.34 (Γ)	2.24 (Γ)	NA
5	0.51	0.49	NA	2.00	NA	2.36 (X)	2.35 (X)	NA

It could be noticed in Figure 10(a) that the PL peaks of sample 4 and 5 have broad FWHM and weak intensity, which is because sample 4 is around the point for bandgap turning from direct to indirect and sample 5 is indirect bandgap material. The PL peaks of those two samples are considered only due to the D-A transition. As shown in Figure 10(a)and Table 1, the PL peak of sample 4 is located at 2.14 eV, whereas the peak of sample 5 is located at 2.00 eV. The strained AlP mole fraction X_{Al-S} is 0.47 for sample 4 and 0.51 for sample 5, the bandgap E_{G-S} at 10 K is around 2.34 eV and 2.36 eV for sample 4 and sample 5 respectively. It means that for sample 4 and 5 the shift direction of the PL emission peak differs from that of the bandgap energy. The analogous composition-dependent red shift of D-A transition PL peak was also reported in beryllium doped $Al_xGa_{1-x}As$ (Morita et al. 1989). In our study, the donor ionization energy (E_D) of sample 5 with X energy valley might be larger than that of sample 4 with Γ energy valley, which could induce the decrease of D-A transition energy for sample 5.

The thermal activation energy and nonradiative recombination mechanism of the samples could be analyzed by the thermal quenching process of PL intensity. We have measured the PL spectra from 10 K to above 100 K at the same excitation power of 13 mW. Figure 13 shows the PL intensity of the D-A transition peak for sample 1-3 as a function of reciprocal temperature (in Log scale), and the intensities are normalized to the maximum intensity of sample 1. The curves show Arrhenius behavior with two different slopes, which could be interpreted in terms of different nonradiative recombination mechanisms respectively prevailing in two distinct temperature ranges. The slopes can give the activation energies for the two nonradiative recombination processes. The line in Figure 13 is the least-squares fit of the experimental data of D-A transition peak using the following equation as a function of temperature T [Lambkin et al. 1994 and Yoon et al. 1997]:

$$I(T) = \frac{\alpha}{1 + C_1 \exp(-E_1/kT) + C_2 \exp(-E_2/kT)} \tag{2}$$

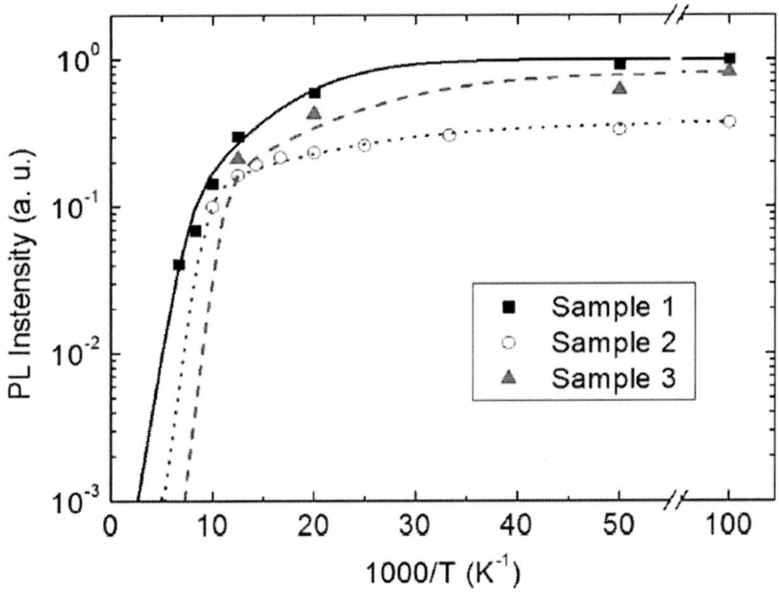

Figure 13. PL intensity of the D-A transition peak for sample 1-3 versus reciprocal temperature from 10 K to above 100 K. The solid, dotted and dashed lines are the least-squares fit of the data to Eq. (2) for sample 1-3 respectively.

where I(T) is PL emission intensity, α is a variable constant, and k is Boltzmann constant, C_1 and $C_2$are the ratios of nonradiative to radiative recombination probabilities for the two loss mechanisms, $E_1$and E_2 are the thermal activation energies. The fitting results of E_1 and E_2 for sample 1-3 have been listed in Table 2. The obtained thermal activation energies are in the range of E_1~8-20 meV and E_2~90-120 meV without significant dependence on lattice mismatch. The thermal activation energy of E_1 is deemed to dominate at lower temperatures and be considered related to silicon donor. On the other hand, the thermal activation energy of E_2 dominates at higher temperatures. The presence of deep acceptors, such as traps or impurities, is considered to be the most likely causes of E_2. Therefore, the value of E_1 plus E_2 should indicate the total D-A pair energy. As listed in Table 2, E_1 plus E_2 is 106 meV, 116 meV and 129 meV respectively for sample 1-3, which is in good agreement with the separation energy between B-B and D-A peaks (E_{BB}-E_{DA}) as listed in Table 2.

Table 2. Thermal activation energies estimated by fitting to the temperature-dependent PL intensity of the D-A transition peaks for sample 1-3 through **Eq. (2)**

Sample No.	E_1 (meV)	E_2 (meV)	E_1+E_2(meV)
Sample 1	17	89	106
Sample 2	8	108	116
Sample 3	10	119	129

2.3. Growth Of Al(Ga)Inp Photodetector Structures

Two kinds of PD epitaxy structures are grown on (100) oriented S. I. GaAs epi-ready substrates, where GaInP and AlInP are applied as absorption layers respectively (denoted as GaInP PD and AlInP PD hereafter). The GaInP PD consists of a 0.9 m n^+ GaAs buffer and bottom contact layer, a 40 nm n^+GaInP back-surface field (BSF) layer, a 1 m n^- GaInP light absorption layer, a 60 nm p^+ GaInP emitter layer, a 30 nm p^+ AlInP window layer and a 0.2 m p^+ GaAs top contact layer, which is quite similar to the top cell structure of the tandem solar cells (e.g.: Karam et al 2001; Yamaguchi 2001). The growth of AlInP PD began with a 1 m n^+ GaInP buffer layer, which was also for the bottom contact, and then a 50 nm n^+ AlInP BSF layer was grown. After that, a 1 m n^- AlInP light absorption layer was grown, which was lightly doped with silicon to below 1×10^{16} cm^{-3} to reach fully depletion at zero bias. Finally, a 150 nm p⁺ AlInP emitter layer and a 150 nm p⁺ GaInP top contact layer were grown.

The grown samples all show shiny surfaces. The structural characteristics of the wafers were measured by XRD as shown in Figure 14 and Figure 15 for GaInP and AlInP PD structures, respectively. From the measurements, the epi-layer of GaInP PD shows a positive mismatch of +9.0110^{-4} to the GaAs substrate, with a FWHM of 28.3 arcsec. For the AlInP PD structure, the peak of AlInP absorption layer shows a positive mismatch of +4.7610^{-4} to the GaAs substrate with a FWHM of 22.5 arcsec as shown in Figure 15, confirming the good epitaxy quality. After that, the grown samples were processed into PD chips for further investigation.

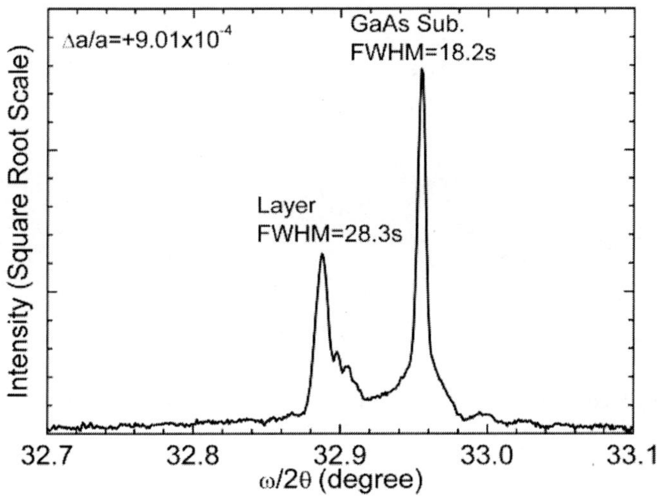

Figure 14. X-ray diffraction rocking curve of the GSMBE grown GaInP PD structure.

3. PERFORMANCES OF SPECIFIC AL(GA)INP PHOTODETECTORS

The grown wafers were processed into mesa type PDs using conventional processing steps. The detector mesas with different diameters were defined by using photolithography and wet etching to the bottom contact layer, and then passivated by plasma enhanced chemical vapor deposition of Si_3N_4. Both the top and bottom contacts were formed using photolithography, evaporation of TiPtAu and lift-off. After an alloy step, the wafers were diced into chips, and then the chips were mounted into transistor outline (TO) packages and wire bonded for further measurements. Various photodiodes tailored for UV-enhanced or blue wavelength range was demonstrated.

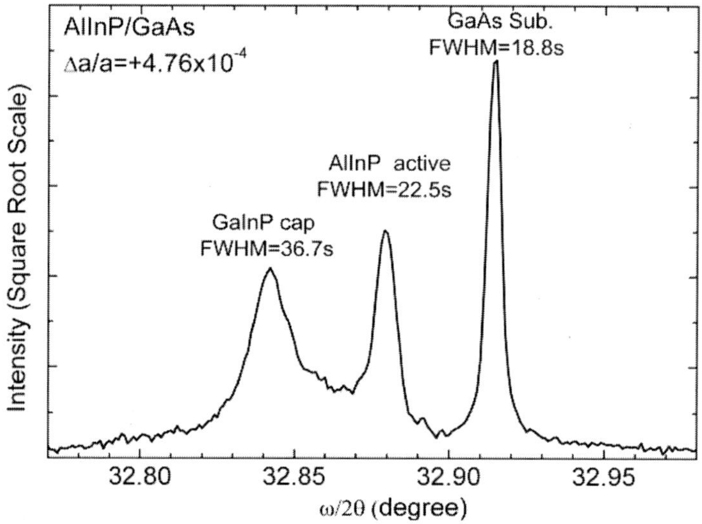

Figure 15. X-ray diffraction rocking curve of the GSMBE grown AlInP PD structure.

3.1. Uv-Enhanced Gainp Photodetector

As concerned in the introduction part of this chapter, many practical and potential applications need PDs sensitive to short wavelength side of visible and extending to UV, but blind to infrared. For GaInP PD, UV light may also be detected besides the visible light and blind to infrared inherently, so we can call it UV-enhanced GaInP PD as demonstrated (Zhang et al. 2005).

Figure 16 (a) shows the schematic of the GaInP detector, (b) is the capacitance-voltage (C-V) characteristics of the PD measured at 1 MHz frequency with a HP4280A C-V meter, the detector sensitive area is

1.0810^{-3} cm^2. The capacitance of this detector is 40.75 pF (37.9 nF/cm^2) at zero bias, including the capacitance of the package and bonding pad. In Figure 16, the carrier concentration along the GaInP absorption layer is also plotted, which is calculated by using $N=(1/q_0\ _rA^2)[C^3/(dC/dV)]$ and $X_d=_0\ _rA/C$ (where $X_d=0$ is at the pn junction interface) from the C-V curve. The electron concentration of $3\sim610^{16}$ cm^{-3} in the GaInP absorption layer is educed, which is in consistent with the quite low silicon doping level during the growth.

The current-voltage (I-V) characteristics, which are directly correlated to the performance of the detector, have been measured using a HP4156A precise semiconductor analyzer over more than 6 orders of magnitude in current range. Figure 17 shows the typical dark I-V characteristics of the detectors at room temperature and reverse bias. The dark current of this GaInP detector is only 340 fA at reverse bias $V_R=50$ mV, the breakdown voltage is great than 5 V. For those GaInP photovoltaic detectors, the output voltage under illumination is much higher than those of Si detectors, the open-circuit voltage of $V_{oc}>1.05$ V is measured as shown in the lower inset of Figure 17. The shunt resistances of the detector at 0 V bias (R_0) are also measured as shown in the upper inset of Figure 17, for this detector R_0 of 120 G ($R_0A=1.310^8$ cm^2) has been reached at room temperature.

The response spectrum of the detector was measured by using a grating monochrometer with lock-in technique, in the measurement a chopped deuterium/halogen combined light source in conjunction with a standard Si detector (with known area and calibrated responsivity data at each wavelength) was used. Figure 18 shows the measured response spectrum of the UV-enhanced GaInP PD at room temperature, the thin solid lines show the quantum efficiency grid and dashed line shows the relative C.I.E. curve of the human eye; the detector was under zero bias. It could be seen that, the response peak of the detector is at 550 nm, with 10% cut-on and cut-off wavelength of 350 nm and 675 nm, respectively. The measured responsivity of the detector at 400, 550 and 650 nm is 0.12, 0.26 and 0.19 A/W, respectively. The quantum efficiency around 500 nm reaches 60% and drops down in both red and blue sides, besides the contact shading and non-optimized AR coating, the reduced efficiency in the red side could be contributed to the insufficient GaInP absorption layer thickness, whereas in blue side the surface recombination may play an important role. In this detector, the infrared response was totally suppressed, the measured infrared suppression ratio is greater than 250 (limited by the measurement system), whereas the UV response is enhanced evidently, this feature is desirable in certain applications. In Figure 18 the Commission

International de l' Eclairage (C.I.E.) curve of human eye was also shown. Notice that the response of this detector covers the C.I.E. curve quite well, so it may also find applications in photometry.

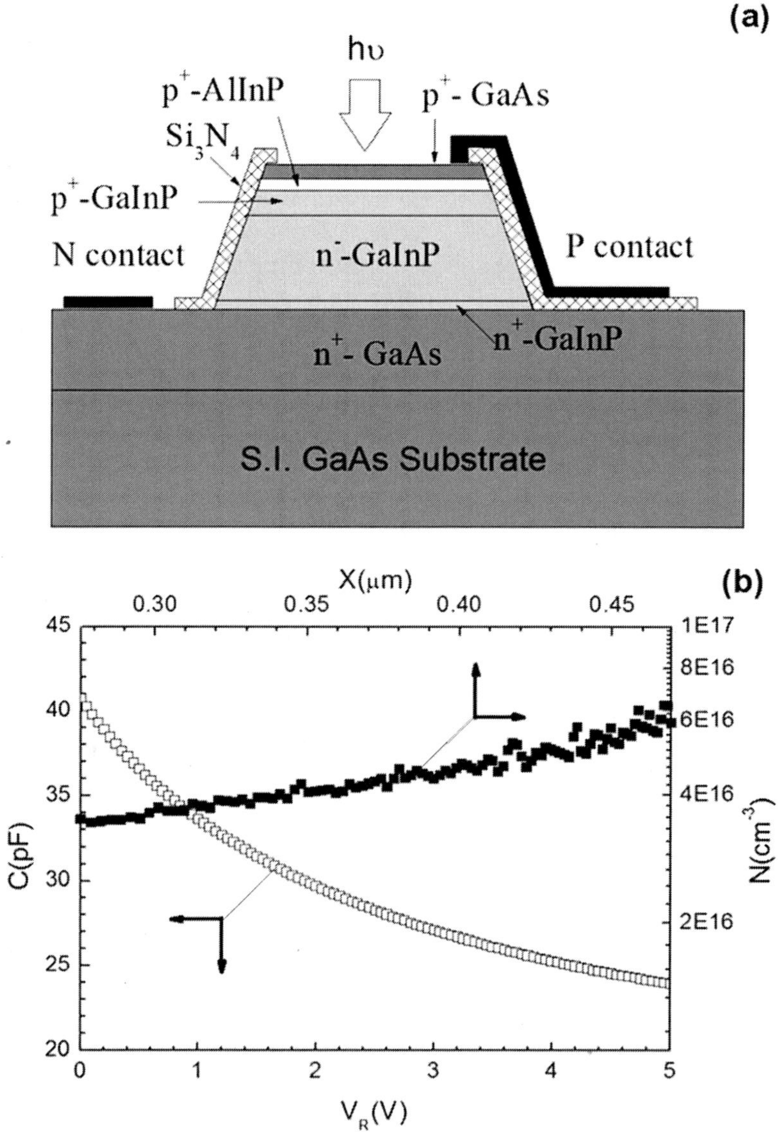

Figure 16. Schematic of the UV enhanced GaInP photodetector (a), and measured C-V characteristics of the PD and calculated carrier concentration profile in the $Ga_{0.51}In_{0.49}P$ light absorption layer.

3.2. Narrow Band Blue Alinp Photodetector

Clear ocean water shows attenuation minimum in the blue (450-490 nm) band with transmission peak around 480 nm, while quite sensitive to particle and pollution (Pope et al. 2000). Therefore for the ocean related remote sensing applications, sensitive PDs in this band with inherent narrow wavelength width should be needed, and photovoltaic type without filter or resonate structure is more preferable. In addition to the traditional Si, II-VI or III-V nitrides PD structures, the AlInP PD adopting AlInP as absorption layer works well in the blue band.

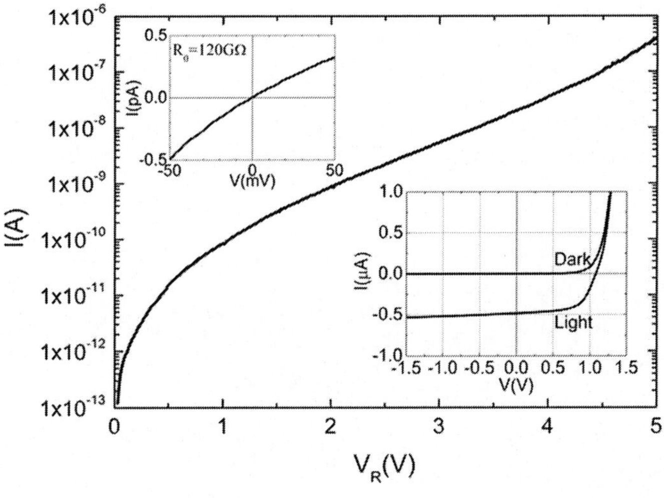

Figure 17. Measured dark I-V characteristics of the UV-enhanced GaInP PD at room temperature and reverse bias; the upper inset shows the I-V characteristics across the zero bias and related shunt resistance of the detector, the low inset shows the I-V characteristics of the detector with and without illumination.

Based on the grown AlInP PD structures mentioned above, devices have been demonstrated with the schematic structure as shown in Figure 19 (Zhang et al. 2010). The typical dark I-V characteristics of the detectors at room temperature and reverse bias are shown in Figure 20; the diameter of the detector is 300 m. It can be seen that the dark current of this AlInP/GaInP detector is only 150 fA at reverse bias V_R=500 mV, and the breakdown voltage is greater than 30 V.

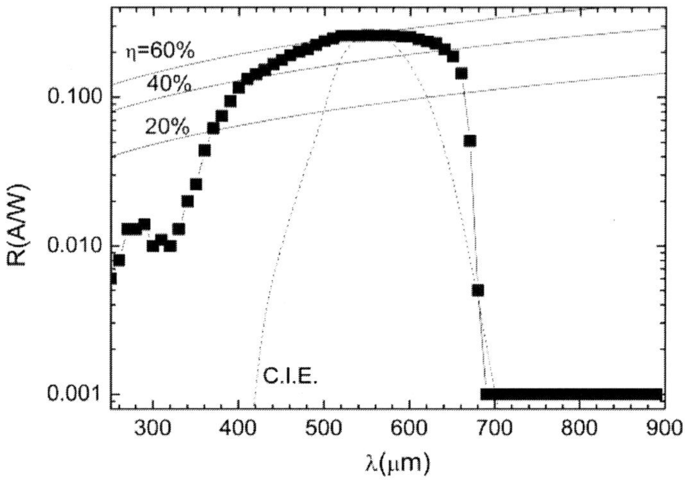

Figure 18. Measured response spectrum of the UV-enhanced GaInP PD; the thin solid lines show the quantum efficiency grid and dashed line shows the relative C.I.E. curve of the human eye.

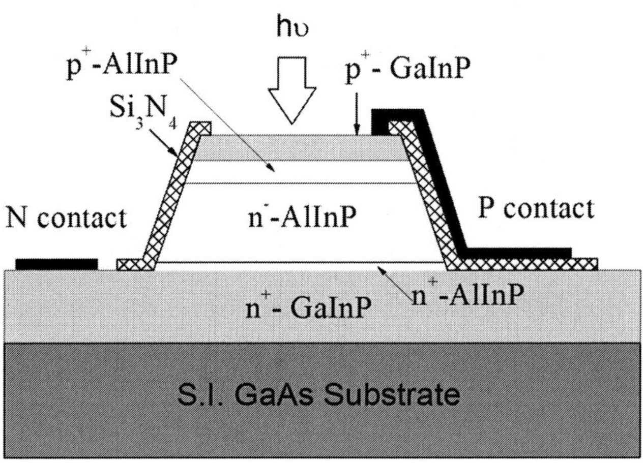

Figure 19. Schematic drawing of the AlInP blue PD.

Because of the lower doping level in AlInP absorption layer, at forward bias <1.78 V the dark current remains below 30 pA, after that the diode turns on. For those GaInP/AlInP photovoltaic detectors, the output voltage under blue light emitting diode illumination is much higher than that of Si detectors, the open-circuit voltage of V_{oc}>1.28 V is measured as shown in the lower inset of Figure 20, the illumination intensity on the detector is about 2 W. The I-V characteristics around zero bias have also been measured as shown in the upper inset of Figure 20, from which the shunt resistances of the detector at 0 V bias (R_0) are deduced. For this detector R_0 of 980 G (R_0A=6.910^8 cm^2) has been reached at room temperature.

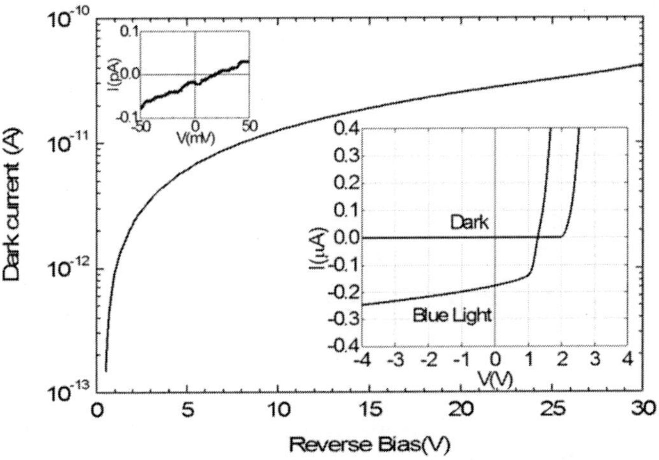

Figure 20. Measured dark I-V characteristics of the $Ga_{0.51}In_{0.49}P$/ $Al_{0.52}In_{0.48}P$/GaAs blue photovoltaic detector at room temperature and reverse bias; the upper inset shows the I-V characteristics around zero bias, the low inset shows the I-V characteristics of the detector in dark and with blue illumination.

The response spectrum of the detector has been measured by using a grating monochrometer adopting lock-in technique. In the measurement an xenon light source chopped at 14 Hz and a calibrated $LiTaO_3$ pyroelectric detector were used. Figure 21 shows the measured responsivity spectrum of the detector under zero bias at room temperature. It could be seen that the detector has a Gaussian alike response spectrum, the peak response wavelength $_p$ is 480 nm with 50% cut-on and cut-off wavelength at 452 nm and 497 nm respectively, corresponding to a relative response wavelength width (FWHM/$_p$) of 9.4%. The measured responsivity of the detector at 480 nm and zero bias is 0.168 A/W, corresponding to an external quantum efficiency of 43.4% for this detector without AR coating. Considering ~30% reflecting loss and ~11% top contact shadowing loss, the peak internal quantum efficiency reaches 70%. In Figure 21 the absorption feature of water is also shown, the response of the detector matches the transmission window quite well.

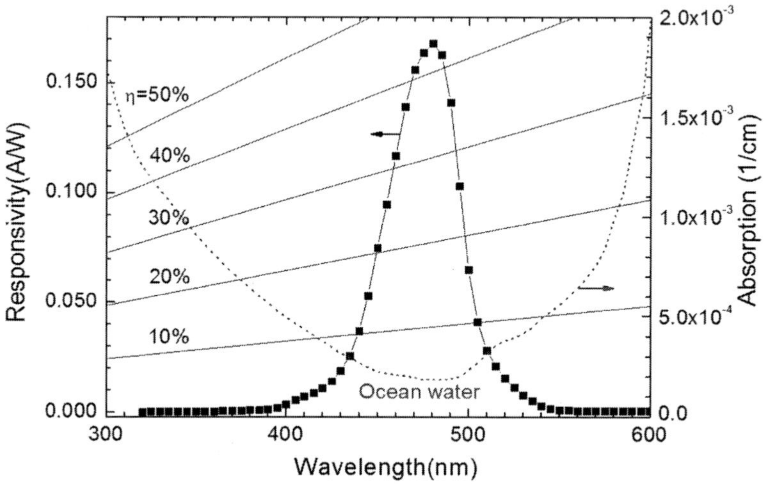

Figure 21. Measured responsivity spectrum of the AlInP blue PD; the thin solid lines show the quantum efficiency grid and dashed line shows the absorption feature of ocean water in this wavelength range.

The response of those detectors is fitly around blue band with intrinsic narrow response wavelength width, which is much lower than the normal value of above 30% for other detectors in this band. This can be attributed to our epitaxy structure design adapting the bandgap features and combination of AlInP with GaInP. At aluminum composition around 0.52 lattice matching to GaAs, AlInP is an indirect bandgap material with bandgap around 2.3 eV corresponding to wavelength about 540 nm, but enters direct bandgap around 2.6 eV (Ishitani et al. 1997; Menoni et al. 1997; Gu et al. 2006). Therefore, the absorption at blue band around 480 nm should be quite strong like those of direct bandgap material. At the same time the response at short wavelength side, especially the UV parts, is restrained by the GaInP cap layer with narrower direct bandgap of about 1.9 eV; furthermore, the response at long wavelength side is also suppressed by the use of a relative thin AlInP absorption layer. The combination of all the effects forms an appropriate response characteristic suitable for blue band detection.

For the same epitaxial structure of the detector, through the etching away of the top p+-GaInP contact layer during the processing to open a light entrance window, the peak response wavelength of the PD blue shifted 10 nm to 470 nm as shown in Figure 22, the peak response increased obviously to 0.287 A/W (Li et al. 2011). The cut-off wavelength at short wavelength side also blue shifted 12 nm to 440 nm, whereas almost unchanged at long wavelength side. The narrow response feature of the PD still retained (increased slightly to about 12%), the dark current still kept at a quite low level ($R_0A=1.710^8 cm^2$), which makes this PD very useful for ocean related blue band applications.

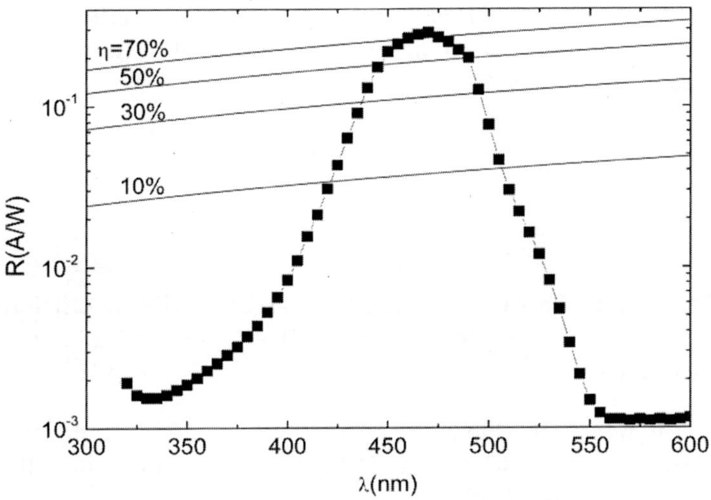

Figure 22. Measured responsivity spectrum of the AlInP PD when etching away of the top p+-GaInP contact layer during the processing.

4. CONCLUSION

In conclusion, based on the analysis of the material issues for the PDs in UV-visible bands suitable for specific applications, the quaternary Al(Ga)InP system lattice matched to GaAs substrate, especially the ternaries GaInP and AlInP, were chosen as our object. Using GSMBE method, the growth techniques and material characteristics of ternaries GaInP and AlInP were investigated in detail (Gu et al. 2006, 2007). The

UV-enhanced AlInP-GaInP-GaAs PD and narrow band AlInP-GaAs blue PD structures have been grown by using GSMBE, and PDs with excellent performances have been demonstrated (Zhang et al. 2005, 2010; Li et al. 2011). The spectral responses of those PDs are tailored by the design of the epitaxial and device structures to specific wavelength band, which are desirable for particular applications where selective response and outstanding performances are expected.

ACKNOWLEDGEMENTS

The authors wish to acknowledge the support of National Basic Research Program of China under grant No.2012CB619200 and Innovative Founding of Shanghai Institute of Microsystem and Information Technology, CAS.

REFERENCES

1. K. Ando, H. Ishikura, Y. Fukunaga, et al. 2002 Highly efficient blue-ultraviolet photodetectors based on II-VI wide-bandgap compound semiconductors. phys. stat. sol. (b) 2292106510711521-3951

2. P. C. Chang, C. H. Chen, S. J. Chang, et al. 2006 High UV/visible rejection contrast AlGaN/GaN MIS photodetectors. Thin Solid Films, 4981-21331360040-6090

3. J. F. Geisz, D. J. Friedman, J. S. Ward, et al. 2008 40.8% efficient inverted triple-junction solar cell with two independently metamorphic junctions. Appl. Phys. Lett. 93121235050003-6951

4. Y. Gu, Y. G. Zhang, H. Li, et al. 2006 Gas source MBE growth and doping characteristics of AlInP on GaAs. Mater. Sci. & Eng. B1311-349530921-5107

5. Y. Gu, Y. G. Zhang, A. Z. Li, et al. 2007 Optical properties of gas source MBE grown AlInP on GaAs. Mater. Sci. & Eng. B1392-32462500921-5107

6. G. Y. Ishitani, E. Nomoto, T. Tanaka, et al. 1997 The energy band alignment of Xc, c, and v points in (Al0.7Ga0.3)0.5In0.5P/AlxIn1-xP heterostructures. J. Appl. Phys. 814176317700021-8979

7. N. H. Karam, R. R. King, M. Haddad, et al. 2001 Recent developments in high-efficiency Ga0.5In0.5P/GaAs/Ge dual- and triple-junction solar cells: steps to next-generation PV cells. Sol. Energy Mater. Sol. Cells 661-44534660927-0248

8. C. Kaspari, M. Zorn, M. Weyers, et al. 2008 Growth parameter optimization of the GaInP/AlGaInP active zone of 635 nm red laser diodes. J. Crystal Growth 31023 517551770022-0248

9. J. D. Lambkin, L. Considine, S. Walsh, et al. 1994Temperature dependence of the photoluminescence intensity of ordered and disordered In0.48Ga0.52P. Appl. Phys. Lett. 65173750003-6951

10. K. H. Lee, P. C. Chang, S. J. Chang, et al. 2012 GaN-based Schottky barrier ultraviolet photodetector with a 5 -pair AlGaN-GaN intermediate layer. phys. stat. sol. (a) 20935795841862-6319

11. C. Li, Y. G. Zhang, Y. Gu, et al. 2011Gas source MBE grown Al0.52In0.48P photovoltaic detector. J. Crystal Growth323 15015030022-0248

12. W. Li, J. Lammasniemi, A. B. Kazantsev, et al. 1998 GaInP/AlInP tunnel junction for GaInP/GaAs tandem solar cells. Electron. Lett. 3444064070013-5194

13. M. Longo, A. Parisini, L. Tarricone, et al. 2001 Photoluminescence investigation of superlattice ordering in organometallic vapour phase epitaxy grown InGaP layers. Mater. Sci. & Eng. B862-31571640921-5107

14. C. S. Menoni, O. Buccafusca, M. C. Marconi, et al. 1997 Effect of indirect -L and -X transfer on the carrier dynamics of InGaP/InAlP multiple quantum wells. Appl. Phys. Lett. 70 1 102 1040003-6951

15. M. Morita, K. Kobayashi, T. Suzuki, et al. 1989 Photoluminescence from Highly Be-Doped AlGaAs Grown by MBE. Jan. J. Appl. Phys. 28 3 553 5540021-4922

16. M. Mosca, J. L. Reverchon, N. Grandjean, et al. 2004 Multilayer (Al, Ga)N structures for solar-blind detection. IEEE J. of Selected Topics in Quantum Electronics1047527580107-7260X

17. H. Murata, Y. Terui, M. Saitoh, et al. 1991 Low threshold current density of 620 nm band MQW-SCH AlGaInP semiconductor lasers with Mg doped AlInP cladding layer. Electron. Lett.27171 56915710013-5194

18. J. S. L. Ong, J. S. . Ng, A. B. Krysa, et al. 2011 Impact Ionization Coefficients in Al0.52In0.48P. IEEE Electron Device Lett. 3211152815300741-3106

19. D. L. Pulfrey, J. J. Kuek, M. P. Leslie, et al. 2001 High UV/solar rejection ratios in GaN/AlGaN/GaN PIN photodiodes. IEEE Trans. on Electron Devices 4834864880018-9383

20. R. M. Pope, A. D. Weidemann, E. S. Fry, 2000 Integrating cavity absorption meter measurements of dissolved substances and suspended particles in ocean water. Dynamics of Atmospheres and Oceans311-43073200377-0265

21. M. Razeghi, A. Roraiski, 1996 Semiconductor ultraviolet detectors. J. Appl. Phys. 7910743374730021-8979

22. L. Rigutti, M. Tchernycheva, A. D. L. Bugallo, et al. 2010 Ultraviolet Photodetector Based on GaN/AlN Quantum Disks in a Single Nanowire. Nano. Lett.108293929431530-6984

23. T. Takamoto, E. Ikeda, H. Kurita, et al. 1997 Over 30% efficient InGaP/GaAs tandem solar cells. Appl. Phys. Lett. 7033813830003-6951

24. I. Vurgaftman, J. R. Meyer, L. R. Ram-Mohan, 2001 Band parameters for III-V compound semiconductors and their alloys. J. Appl. Phys. 8911581558750021-8979

25. M. Yamaguchi, 2001 Radiation-resistant solar cells for space use. Sol. Energy Mater. Sol. Cells, 68131350927-0248

26. I. T. Yoon, B. S. Jeong, H. L. Park, 1997 Zn diffusion of In0.5Ga0.5P investigated by photoluminescence measurements. Thin Solid Films, 3001-22842880040-6090

27. S. F. Yoon, K. W. Mah, H. Q. Zheng, 1999 Transport and photoluminescence of silicon-doped GaInP grown by a valved phosphorus cracker cell in solid source molecular beam epitaxy. J. Appl. Phys., 8510737473790021-8979

28. E. Zacks, A. Halperin, 1972 Dependence of the peak energy of the pair-photoluminescence band on excitation intensity. Phys. Rev. B 68307230751098-0121

29. T. C. Zhang, Y. Guo, Z. X. Mei, et al. 2009 Visible-blind ultraviolet photodetector based on double heterojunction of n-ZnO/insulator-MgO/p-Si. Appl. Phys. Lett. 94111135080003-6951

30. Y. G. Zhang, A. Z. Li, A. G. Milnes, 1997 Metal-semiconductor-metal ultraviolet photodetectors using 6H-SiC. IEEE Photon. Technol. Lett. 933633641041-1135

31. Y. G. Zhang, Y. Gu, C. Zhu, et al. 2005 AlInP-GaInP-GaAs UV-enhanced photovoltaic detectors grown by gas source MBE. IEEE Photon. Technol. Lett. 176126512671041-1135

32. Y. G. Zhang, C. Li, Y. Gu, et al. 2010 GaInP-AlInP-GaAs blue photovoltaic detectors with narrow response wavelength width. IEEE Photon. Technol. Lett. 22129449461041-1135

9

INFRARED PHOTODIODES ON II-VI AND III-V NARROW-GAP SEMICONDUCTORS

Volodymyr Tetyorkin[1], Andriy Sukach[1] and Andriy Tkachuk[2]

[1] V. Lashkaryov Institute of Semiconductor Physics NAS of Ukraine, Ukraine

[2] V. Vinnichenko State Pedagogical University, Ukraine

1. INTRODUCTION

During the last two decades HgCdTe, InSb and InAs infrared (IR) photodiodes have developed rapidly for utilization in second generation thermal-imaging systems. Obviously, they are regarded as the most important candidates for development of third generation systems as well. Despite this fact many problems still exist in manufacturing technology as well as in understanding of physical phenomena in materials and photodiodes. As a result, threshold parameters of commercially available IR photodiodes are far from the values predicted theoretically.

The concept of band gap engineering have found numerous applications in the fabrication IR devices on II-VI and V III-V semiconductors. For instance, the most important advantage of HgCdTe ternary alloy is ability to tune its energy band gap in wide range. The spectral cutoff of HgCdTe photodiodes can be tailored by adjusting the HgCdTe alloy composition over the 1-30 mm range. Further application of this concept in technology of IR detectors is closely connected with development of GaAs/AlGaAs multiple quantum well detectors and InAs/GaInSb type-II superlattice photodiodes.

To implement the concept of defect engineering, grown-in and process-induced defects must be minimized and passivated or eliminated. Defects in narrow-gap semiconductors are easily introduced either intentionally or unintentionally during crystal growth, sample treatment and device processing. There are also evidences that these defects are electrically active. So, for controlling parameters and characteristics of infrared photodiodes on narrow-gap semiconductors through defect engineering, it is essential to understand physical properties of defects, mechanisms of their interaction and temporal evolution. Electronic properties of native defects and foreign impurities in narrow-gap semiconductors have been of great importance for several decades. As a result of intensive investigations, the primary native defects and the mechanisms of their formation were elucidated. Doping effect of different impurities has been also recognized. This allows to develop effective methods for controlling the carrier concentration and type of conductivity in intentionally undoped and doped materials. To some extent the carrier lifetime can be controlled by extrinsic doping. However, in many cases electronic states of defects in these semiconductors are still to be unknown and further investigations are needed.

For controlling properties of semiconductors through defect engineering, it is essential to understand the mechanisms of interaction between point and extended defects (dislocations), as well as to understand their effect on device characteristics. This task seems to be important since alternative substrates (Si, GaAs, sapphire) are widely used in epitaxial technology of IR photodiodes. These substrates are very attractive because they are less expensive and available in large area wafers. The coupling of the Si substrates with Si read-out integrated circuit allows fabrication of very large focal-plane arrays (FPA). Due to the large lattice mismatch between HgCdTe, InSb and InAs and alternative substrates, photodiodes on their base suffer from the high density of dislocations (typically of the order of 10^6 cm^{-2}). For instance, these defect densities seriously limit application of HgCdTe epitaxial layers for manufacture of high-performance photodiodes for the LWIR and VLWIR spectral regions. The use of buffer layers, temperature cycling and hydrogen passivation is expected to be useful for reduction of the density of dislocations and weakening their effect on the device performance. However, none of these methods has yet been proven to be practical.

A number of physical properties of HgCdTe, such as direct energy gap, ability to obtain both low and high carrier concentrations, high mobility of electrons and holes, low dielectric constant and extremely small change of lattice constant with composition, makes it possible to grow high quality layers and heterostructures. As a consequence, high-performance HgCdTe photodiodes on mead-wavelength, long-wavelength and very long-wavelength IR regions (MWIR, LWIR and VLWIR) have been developed. The main drawbacks of HgCdTe are technological disadvantages of this material. The most important is a weak Hg–Te bond, which results in bulk, surface and interface instabilities. Uniformity and yield are still issues especially in photodiodes on the LWIR and VLWIR spectral regions. InSb photodiodes on the MWIR spectral region have comparable performance with photodiodes made of HgCdTe.

Initially, the IR photodiodes were prepared by diffusion, ion implantation or other techniques which allow preparation of homojunction structures. In these photodiodes, the concentration of carriers in the lightly doped 'base' region was strongly controlled. With development of epitaxial techniques, homojunction photodiodes were replaced by heterojunction ones. In a heterojunction photodiode the 'base' region are introduced between a wide-gap substrate and a capping layer with wider band gap. Thus, the influence of surface recombinations on the photodiodes performance is weakened. It seems that the most successful application of the band gap and defect engineering concepts in technology of IR photodiodes is development of double-layer heterojunction photodiodes.

The main objective of this article is to outline the basic properties of point and extended defects, their effect on physical properties and threshold parameters of infrared photodiodes based on HgCdTe, InAs and InSb narrow-gap semiconductors. This article is divided into two parts. The first part is dedicated to technological steps (crystal growth, thermal annealing, junction formation) closely connected with defects forming in materials and devices. In the second part original results are analyzed with emphasize on possible participation of dislocations and point defects in the carrier transport mechanisms and recombination processes in the photodiodes.

2. NATIVE DEFECTS AND IMPURITIES IN HGCDTE, INAS AND INSB

The defect structure of narrow-gap $Hg_{1-x}Cd_xTe$ (hereinafter – HgCdTe) compounds was intensively investigated both theoretically and experimentally over the past fifty years. The current status of defect states in these semiconductors are reviewed in numerous papers and monographs (see, e.g., Capper and Garland, 2011; Chu and Sher, 2010). HgCdTe crystalline materials are always grown with large deviation from stoichiometry. The equilibrium existence region in HgCdTe (x~0.2) is shown to be completely on the Te-rich side (Schaake, 1985). Thus, the most important type of native point defects in undoped materials are Hg vacancies. Residual impurities, Hg interstitials, dislocations and Te precipitates were also observed in as-grown materials. All these defects can exist in neutral or ionized states. Their important characteristics, such as donor or acceptor type, ionization energy, density, spatial location and temporal variation were investigated.

The ab initio calculation of the formation energies of native point defects in HgCdTe (x~0.2) has been made by Berding with co-authors (Berding, 1994, 1995, 2011). The most reliable calculations are based on the local-density approximation to the density functional theory. The calculations predict the mercury vacancy and the tellurium antisite Te_{Hg} as the dominant defects in the material grown under tellurium-rich conditions. The concentration of native point defects was calculated as a function of Hg pressure using quasichemical formalism. In the the calculation all defects were assumed to be equilibrated at the temperature 500 0C and the defect concentrations were assumed to be frozen in. The 500 ^{0}C temperature is typical of LPE growth. In the calculations the Hg vacancy and tellurium antisite Te_{Hg} are classified as acceptor and donor defects, respectively. The calculation also predicts that the concentration of Te_{Hg} is comparable with the mercury vacancies concentration. At the same time, as was pointed by Berding, to date there is no experimental confirmation of the presence of the tellurium antisite defects in HgCdTe in so large amounts. Also, the calculated concentration of Hg interstitial was found to be too low to explain experimental data on the self diffusion (Berding, 2011). Experimentally the defect structure in undoped HgCdTe (x~0.2) was investigated by Vydyanath and Schaake. It has been shown that the dominant native defects are doubly ionized acceptors associated with Hg vacancies (Vydyanath, 1981; Schaake, 1985). This result is in accordance with the theoretical prediction. Defects in doped HgCdTe have been reviewed by Shaw and Capper (Shaw and Capper, 2011).

Native defects in HgCdTe, including dislocations and defect complexes, can act as Shockley-Read-Hall (SRH) centers due to their effect on the carrier lifetime. There is a large literature concerning the links of deep defects to the carrier lifetime in HgCdTe (Capper, 1991; Sher, 1991; Capper, 2011;Cheung, 1985, Chu and Sher, 2010). Clear evidence of SRH centers was provided by deep level transient spectroscopy, admittance spectroscopy, thermally stimulated current and optical modulation spectroscopy (Polla, 1981, 1981a, 1981b, 1982; Jones, 1982; Schaake, 1983; Mroczkowski, 1981). The centers located at near midgap seems to be common to p-type Hg_1CdTe, where the doping is due to mercury vacancies. Summary of impurity and native defect levels experimentally observed in HgCdTe has been done by Litter et. al. (Litter, 1990). The main results of experimental findings are as follows: (i) shallow acceptor-like levels have activation energies between 2 and 20 meV; deep levels have energies $0.25E_g$, $0.5E_g$ and $0.75E_g$ above the valence band edge. The concentration of donor-like deep centers reported by Polla and co-authors was ranged from approximately $0.1N_A$ to $10N_A$, where N_A is the shallow acceptor concentration. The values of the cross section for electrons and holes were in the range 10^{-15}-10^{-16} cm^2 and 10^{-17}-10^{-18} cm^2, respectively. The origin of the SRH centres in HgCdTe is still not clear. The vacancy-doped materials with approximately the same carrier concentration, but manufactured at different temperature conditions, may exhibit different lifetimes. At the same time, the correlation has been found between the SRH recombination centre densities and the Hg vacancies concentration (Capper, 2011).

The behavior of extrinsic defects in HgCdTe is important from several reasons. Manufacture of photodiodes with improved characteristics requires intentionally doped materials with controllable concentrations of acceptors and donors instead of vacancies-doped material. This is caused by several reasons such as instability of the vacancy-doped material and low ability to control the concentration of free carriers. Also, Hg vacancies or complexes with their participation may be responsible not only for the carrier lifetime, but also they can enhance the trap-assisted tunneling, giving rise to excess dark current in infrared photodiodes. Shaw and Capper (Shaw and Capper, 2011) provides a complete summary of the work on dopants in bulk material and epitaxial layers. Indium and iodine are most frequently used as a well-controlled donors for preparation of n-type bulk material and epitaxial layers. Both are incorporated as shallow single donors occupying metal and tellurium lattice sites (In and I, respectively). Indium has moderately high diffusivity whereas diffusivity of iodine is rather low. Group V elements have low diffusivity and hence are ideal as p-type dopants. Approximately 100% activity is found for In and I concentrations up to ~10^{18} cm^{-3}. Vydyanath (1991) argued that group V elements (P, As, Sb, Bi) are incorporated as donors under Te-rich (Hg-deficient) conditions of growth. Under Hg saturated anneal at 500 ^{0}C there is enough energy to move these elements from metal lattice cites to Te or interstitial sites. The group V impurities if occupy Te sites act as shallow acceptors.

Shallow impurities are known to determine the concentration of free carriers in semiconductors. The ionization energy of a hydrogen-like donors in HgCdTe were calculated as a function of composition and concentration of defects (Capper, 2011). Due to low effective masses of electrons this energy is too small to be detected experimentally at 77 K (e.g., the ionization energy is 0.30 (0.85) meV for the composition x=0.2 (0.3). The ionization energy for acceptors depends on the material composition, concentration of acceptors and degree of compensation. The calculated energy for the shallow acceptors is 11(14) meV for x=0.20 (0.3).

Electronic properties of extended defects in HgCdTe, InSb and InAs are less investigated in comparison with point defects. It is known that II-VI semiconductors are more ionically bonded as III-V covalent semiconductors. As a result, they can be easily plastically deformed at room and lower temperatures (Holt and Yacobi, 2007). The photoplastic effect, discovered by Osipiyan and Savchenko (1968), is conclusive proof that dislocations in II-VI materials are electrically charged and that charge is largely electronically determined. The dislocation core contains broken bonds so the dislocation line may generally be charged negatively. The broken bonds are chemically reactive and electrostatic interaction between the charged dislocation lines and ionized point defects and short-range chemical bonding effects may occur. This tends to reduce (neutralize or passivate) the electrical effects of dislocations especially in the ionically bonded II-VI compounds. The charge states of the dislocation can be altered illumination and other means of carrier injection and this can change dislocation line charges as well as dislocation mobility. Yonenaga (1998) compared the dynamics of dislocations in InAs with those in other semiconductors, including narrow-gap and wide-gap II-VI compounds. It has been shown that the activation energies for dislocation motion depend linearly on the band-gap with an apparent distinction between different types of semiconductors. The activation energy is lower in ionically bonded II-VI semiconductors compare to III-Vs. Thus, dislocations would be expected to be the most mobile in II-VI narrow-gap semiconductors. The long-range electrostatic interaction strengthens the attraction of dislocations to their Cottrell impurity atmospheres and strengthens the pinning effect opposing dislocation motion.

Dislocations can also act as SRH centers in HgCdTe and III-V narrow-gap semiconductors. The electrical activity of dislocations can be attributed to Cottrell atmosphere, or to dangling bonds in the dislocation core. The ability of the strain fields of dislocations in HgCdTe to capture significant amounts of impurities has been investigated in the early work by Schaake (Schaake, 1983). It has been argued that the origin of the electrical activity is dangling bond states on the dislocations rather that the impurities. The high Peierls stress observed in II-VI compound semiconductors supports this argument. This stress has been attributed to the ability of dangling bonds to heal themselves along the dislocation core. The reduced lifetime and mobility caused by dislocations in n- and p-type HgCdTe has been pointed out in several papers (Lopes, 1993; Yamamoto, 1985; Shin, 1991). This reduction has been ascribed to dangling bonds which provide SRH centers. Dislocations also affect the dark currents in LWIR photodiodes operating at low temperatures (<77 K) because they are believed to produce mid-gap states in the band gap (Arias, 1989; Tregilgas, 1988). The decrease of the differential resistance-area product at zero bias voltage, R_0A, in the presence of high dislocations densities has been reported by Johnson et al. (Johnson, 1992). As the temperature decreases the effect of dislocations was found to be more significant. At 77 K the decrease of R_0A begins at the dislocation density of the order of 10^6 cm^{-2}, whereas at 40 K it is affected by the presence of one or more dislocations. The scatter in the R_0A data may be associated with the presence of pairs of 'interacting' dislocations, which may be more effective in reducing the R_0A than individual dislocations. The excess current in photodiodes caused by dislocations may be the source of $1/f$ noise.

The effect of misfit dislocations on dark currents in high temperature MOCVD HgCdTe infrared heterostructure photodiodes has been investigated by Jóźwikowska et al. (Jóźwikowska, 2004). It was shown that the most effective current transport mechanism at high temperature in HgCdTe heterostructures is the trap-assistant tunneling. In the photodiodes operated at 240 K, this mechanism is predominant at bias voltage that not exceeded 0.1 V. The best fit of experimental data with theoretical predictions for the zero bias differential resistance versus temperature has been obtained for rather high dislocation densities in the volume of HgCdTe layer $\sim 5\ 10^7$ cm^{-2}. To a certain extent electrical activity of dislocation can be reduced by passivation (Boieriu, 2006). It has been showed that incorporation of H in In-doped HgCdTe ($x = 0.2$) epilayers, through exposure to an electron cyclotron resonance

(ECR) H plasma, the lifetime increases by a factor of 10. The increase was attributed to H passivation of the dangling bonds and is only effective for high dislocation densities (~10^7 cm^{-2}).

The early studies of native defects and impurities in InSb and InAs have been summarized by Milnes (Milnes, 1973). The energy of shallow impurities in InAs and InSb was calculated with extension of effective mass theory (Baldereschi and Lipari, 1974). Several deep donors of undetermined origin have been observed in InSb. Copper atoms segregated at dislocations in InAs are apparently electrically inactive, but on heating they diffuse away and can be frozen into the lattice as electrically active centers. They return to the dislocations during the low-temperature anneal. A similar effect was observed with Cu in InSb. A number of undetermined deep defects was found in InSb and InAs (Madelung, 2003).

The nature of intrinsic point defects in InAs single crystals has been studied by several groups (Bublik, 1977, 1979, 1979a; Karataev, 1977; Mahony and Maseher, 1977). In specially undoped InAs single crystals grown by direct solidification and Czochralski methods precision measurement of density and lattice constant has been made in order to determine the type and concentration of interstitial atoms and vacancies depending on the content of arsenic in InAs melt. The difference in concentrations of V_{As} and As_i for InAs was found to be of the order of 3 10^{18} cm^{-3}. For InAs grown from the melt of stoichiometric composition, this difference does not exceed $1 \cdot 10^{17}$ cm^{-3}. It was also established the effect of point defects on structural and recombination properties of InAs single crystals. It is shown that InAs crystals were n-type conductivity and electron concentration increased from $1.4 \cdot 10^{16}$ to $2.5 \cdot 10^{16}$ cm^{-3} (T = 77 K) with increasing of As content in the growth melt. It has been concluded that the concentration of electrons is determined by intrinsic defects and complexes composed of native defects (vacancies) and background impurities. The effect of annealing on electrical properties of undoped indium arsenide has been investigated by Karataev (1977). The annealing was made in the temperature range 300-900 °C for 1-100 h. It was found that the annealing increases the concentration of electrons. For example, for the annealing temperature ~900 °C the concentration of electrons increased from $2 \cdot 10^{16}$ cm^{-3} to $2 \cdot 10^{17}$ cm^{-3}.

Table 1. Parameters of impurities and structural defects in InAs

Impurity, structural defect	Type	Ionization energy, meV	Concentration, cm^{-3}	Reference
As divacancy+ residual impurity	neutral complex	-	$(7 \pm 2) \cdot 10^{16}$	Mahony, J. and Maseher, P., 1977
As monovacancy+ residual impurity	neutral complex	-	$\sim 1 \cdot 10^{17}$	Mahony, J. and Maseher, P., 1977
V_{In}	A_1	E_V+100	-	Mahony, J. and Maseher, P., 1977
V_{In}	A_2	E_V+130	-	Mahony, J. and Maseher, P., 1977
V_{In}	A_3	E_V+230	-	Mahony, J. and Maseher, P., 1977
V_{As}	D	$E_C+0.03E_g$	-	Bynin, M.A. and Matveev, Yu.A., 1985
V_{As}	D	$E_C-E_g/4$	-	Bynin, M.A. and Matveev, Yu.A., 1985
Cr	D	E_C-160	-	Omel'yanovskii,1975; Balagurov, 1977; Piltnikas,1982; Adomaytis,1984
S, Se, Te	D	$E_C-(2-3)$	-	Voronina, 1999
Mn	A	$E_V+(28\text{-}30)$	-	Adrianov, 1977; 26 Gheorghitse,1989
Zn	A	E_V+10	-	Kesamanly, 1968
Zn	A	E_V+25	-	Guseinov, 1969; Guseinov, 1997
Cd	A	E_V+20	-	Galkina, 1966
Cd	A	E_V+11	-	Iglitsyn, 1968
Mg	A	-	-	Voronina, 2004
Cu	D	$E_C-< 7$	-	Karataev, 1977
Pb	N	-	-	Baranov, 1992; Baranov, 1993
Be	A	$E_V+< 7$	-	Lin, 1997; Astahov, 1992; Dobbelaere, 1992
Ge (amphoteric)	D (Ge sub.In)	$E_C-< 7$	-	Guseva, 1974; Guseva, 1975
Ge (amphoteric)	A (Ge sub. As)	E_V+14	-	Guseva, 1974; Guseva, 1975
Si (amphoteric)	D (Si sub. In)	$E_C-< 7$	-	Guseva, 1974; Guseva, 1975
Si (amphoteric)	A (Si sub. As)	E_V+20	-	Guseva, 1974; Guseva, 1975
Structural defect	-	110±20	$(0.5\div10) \cdot 10^{14}$	Fomin, 1984
-	-	150±20	$\leq 5 \cdot 10^{14}$	Fomin, 1984
-	-	~ 220	$(1\div3) \cdot 10^{15}$	Ilyenkov, 1992
-	-	$\sim E_g/2$ (77 K)	$\sim 10^{17}$	Kornyushkin, 1996
-	D	$E_C-(100\text{-}200)$	-	Baranov, 1992
-	D	$E_C-(10\text{-}20)$	-	Baranov, 1992
-	A	E_V+50	-	Baranov, 1992
-	D	E_C-15	-	Baranov, 1993
-	A	E_V+50	-	Baranov, 1993
-	D	$E_C-(20\text{-}30)$	-	Voronina, 1999
-	D	$E_C-(90\text{-}100)$	-	Voronina, 1999
-	A	E_V+10	-	Voronina, 1999a
-	A	E_V+20	-	Voronina, 1999a
-	A	E_V+30	-	Voronina, 1999a
-	A	E_V+65	-	Voronina, 1999a
-	A	E_V+20	-	Zotova, 1975
-	A	E_V+35	-	Zotova, 1975
-	A	E_V+35	-	Allaberenov, 1970
-	A	E_V+35	-	Esina, 1985
Residual impurity	D	E_C-2	-	Baranov, 1992
Dislocation (possibly)	A	$E_V+(45\text{-}50)$	-	Anisimova, 1969

Table 2. Parameters of impurities and structural defects in InSb[a,b]

Impurity, structural defect	Type	Ionization energy, meV	Concentration, cm^{-3}	Reference
Zn	A	$E_V + 8$	-	Ismailov, 1969
Zn	A	$E_V + 10$	-	Pehek and Levinstein, 1965
Ge	A	E_V+ (16-19)	-	Ismailov, 1969
Ag_1	A_1	E_V+27	-	Pehek and Levinstein, 1965
Ag_2	A_2	E_V+50	-	Pehek and Levinstein, 1965
Au	A	E_V+43	-	Pehek and Levinstein, 1965
Cu	A	E_V+ (57-64)	-	Valyashko, 1975
Cu	A	E_V+(23-27)		Kevorkov, 1980
Cu, Ag, Au	A	E_V+67	$3\cdot10^{14}$	Korotin, 1976
Structural defect	D	E_C–48	-	Valyashko, 1975
-	A	E_V+ (7-9)	$\leq 6\cdot10^{14}$	Kevorkov, 1980
-	A	E_V+ (5-6)	-	Nasledov, 1962
-	D	E_C– (7-8)	-	Nasledov, 1962
-	D	E_C–55	$\sim 8\cdot10^{13}$	Laff and Fan, 1961
-	D	E_C–60	10^{11} - 10^{13}	Sipovskaya and Smetannikova, 1984
-	A	E_V+ (50-60)	$\sim 10^{14}$	Zitter, 1959
-	D	E_C- (65 ± 5)	$(1.3 - 7.5)\cdot10^{13}$	Golovanov, 1973
-	A	E_V+ (20-30)	$1\cdot10^{15}$	Guseinov, 1971
-	D	E_C–110	-	Valyashko, 1975
-	A	E_V+ (150-170)	$(0.3\text{-}1.0)\ 10^{12}$	Tsitsin, 1975
-	D	E_C– (50-70)	-	Trifonov, 1971
-	A	E_V+ 120	-	Golovanov, and Oding, 1969
-	D	E_C– (110-120)	$< 7\cdot10^{13}$	Nasledov, 1962
-	D	E_C–120	$\sim 8\cdot10^{13}$	Laff and Fan, 1961
-	D	E_C–140	$\sim 2\cdot10^{14}$	Volkov and Golovanov, 1967
-	D	E_C–110	$\sim 3\cdot10^{13}$,	Hollis, 1967
-	A	E_V+ (60-63)	-	Egemberdieva, 1982
-	A	E_V+ (69-73)	-	Egemberdieva, 1982
-	A	E_V+ (80-82)	-	Egemberdieva, 1982
-	D	E_V+ (110-113)	-	Egemberdieva, 1982
-	D	E_V+ (131-135)	-	Egemberdieva, 1982
-	D	E_V+ (140-142)	-	Egemberdieva, 1982
-	D	E_C–130	10^{13} - $2\cdot10^{15}$	Sipovskaya and Smetannikova, 1984
-	-	E_V+ (120±30)	$(0.7 - 2.3)\cdot10^{15}$	Golovanov, 1973
-	-	E_C–130	$\sim 1.2\cdot10^{14}$	Korotin, 1976
-	-	E_C–220	$(2 - 5)\cdot10^{12}$	Korotin, 1976
-	-	E_C–110	$2\cdot10^{14}$	Guseinov, 1971
-	A	E_V+68	$(3 - 4)\cdot10^{13}$	Shepelina and Novototsky-Vlasov, 1992
-	D	E_V+99	$(6 - 8)\cdot10^{13}$	Shepelina and Novototsky-Vlasov, 1992
-	D	E_V+132	$(2 - 3)\cdot10^{13}$	Shepelina and Novototsky-Vlasov, 1992
In_i+O_2	A	E_C-50		Zaitov, 1981

a. in Table 1 and 2 residual impurities and structural defects have unknown identity;

b. doping behavior is indicated by A, acceptor, D, donor; A_1,A_2 and A_3 means singly, doubly and triple ionized acceptor.

The greatest effect of the anneal was observed in InAs, grown from the melt with an excess of indium or arsenic atoms. Dislocations with densities from 10^3 to 10^5 cm^{-2} were found to contribute to the effective change in the concentration of electrons at the annealing process. Investigation of defect complexes in InAs using positron annihilation has been presented by Mahony and Maseher (1977). The neutral complexes with the concentration of $(7\pm2)\cdot10^{16}$ cm^{-3} has been found. Moreover, these complexes are composed of two vacancies at high temperatures whereas at low temperatures only one vacancy is participated in a complex. Both complexes are stable up to temperatures ~850 ºC. Theoretical calculations of energy levels related with In and As vacancies was also carried out (Mahony and Maseher, 1977). It is shown that In and As vacancies are acceptors and donors, respectively. Growing the high resistivity InAs LPE films doped with chromium and study its electrical and photovoltaic properties was also reported (Omel'yanovskii,1975; Balagurov, 1977; Plitnikas,1982; Adomaytis, 1984). Based on investigations of deep centers in epitaxial layers of InAs, the development of high-quality MOS-structures with a density of surface states ≤2 10^{10} cm^{-2} eV^{-1} has been realized. Linear and matrix IR photodetector hybrid assemblies with the temperature resolution of 4 - 8 mK were prepared (Kuryshev, 2001, 2009).

The deviation from the stoichiometry in InSb has been investigated by Abaeva et al. (Abaeva, 1987). The authors make a precise determination of the temperature dependence of the InSb lattice constant by Bond's method. It is shown that In and Sb vacancies are the principal intrinsic point defects that determine the deviation from stoichiometry in InSb and reduce the lattice constant. The mid-gap defect states were observed in InSb bulk crystals (Ehemberdyeva, 1982; Shepelyna, 1992). Their energy does not depend on the chemical nature of dopant and the doping level. Defect complex composed of indium and oxygen atoms with energy E_C-50 meV has been found by Zaitov et al. (Zaitov, 1981). The low temperature (~200 ^{0}C) annealing in an atmosphere of inert gas or saturated vapor of lead results in 5-10 times increasing of the carrier lifetime in n-InSb (Tsitsina,1975; Strelnikova, 1993). A model of the conductivity type conversion with participation of fluorine has been proposed (Blaut-Blachev,1979; Kevorkov, 1980). Some parameters of impurities and deep defects in InAs and InSb are summarized in Table. 1 and 2.

3. DEFECT-RELATED MANUFACTURING PROCESSES

3.1. Growth Techniques

Various methods have been proposed for growing II-VI and III-V semiconductor crystals. These methods can be roughly classified into the melt growth, solution and vapor growth methods. Each of the proposed growth methods has its own advantages and disadvantages. For growing II-VI semiconductor compounds there exist a problem of high dissociation pressure which leads to evaporation of a volatile component and hence deviation from stoichiometry. The deviation from stoichiometry introduces lattice defects since there is a deficiency of one component. If there is a large deviation from stoichiometry, vacancies can combine with the residual impurities or with the impurities which are later introduced into the crystal during a subsequent process. The number of defects is hard to control in this case. Therefore, from the point of perfect crystallography, it is desirable to grow crystals at temperatures as low as possible. This requirement is especially important for HgCdTe alloys, since they are decomposing solids. Also, the liquidus-solidus temperatures of the mercury telluride and cadmium telluride compounds are different, which causes their segregation as the alloy is frozen from a melt. The resultant variation of the mole fraction of each compounds results in a consequent variation in energy gap as well as electrical and optical properties throughout the material. In contrast to HgCdTe, the III-V compounds (InSb and InAs) melt congruently, i.e. a liquid and solid having identical compositions are in equilibrium at the melting point. Thus, they can be grown directly from the melt, a process commonly used to grow large boules of InAs and InSb.

Further progress in IR photodiode technology is connected with epitaxial layers. Epitaxial techniques offer the possibility of growing large area layers with good depth and lateral homogeneity, abrupt composition and doping profiles, which can be configured to improve the performance of photodiodes. The commonly used methods for preparation of epitaxial films are liquid phase epitaxy (LPE), molecular beam epitaxy (MBE), metal-organic vapor phase epitaxy (MOVPE), and metal-organic chemical vapor epitaxy (MOCVD). LPE method is a near-thermodynamic equilibrium growth technology. It has the advantages of relatively simple process, high utilization rate of the source material, high crystalline quality of the epitaxial films and fast growing. The weakness of LEP is that it cannot be used for precision controlled growth of very thin films of nano-scale. In other words, it is not applicable to the growth of superlattices or quantum-well devices and other complex micro-structure materials. In addition, the morphology of materials grown by LPE is usually worse than that grown by MOCVD or MBE. In these techniques epitaxial growth is performed at low temperatures, which makes it possible to reduce the native defects density. Obviously, for epilayers the choice of substrate is decisive for minimizing effect of misfit dislocations and interdiffusion at the epilayer/substrate heterojunction. This problem is especially important for InSb and InAs because of the lack of lattice-matched III-V wide-gap semiconductors which may be used as substrates. To overcome this difficulty, the use of alternative substrates (Si, GaAs, supphire) attracts great attention. Successful implementation of epitaxy of narrow-gap semiconductors on Si substrates can directly lead to possibility of realization of multicolor, monolithic focal plane arrays.

3.1.1. Growth Of Hgcdte

Several historical reviews of the development of bulk HgCdTe have been published (Maier and Hesse, 1980; Capper, 1994; Capper, 1997). Bulk growth of HgCdTe is rather difficult due to the high vapor pressure of Hg at the crystal melting point (about 950 ^{0}C), caused by weak Hg-Te bond in a crystal. For bulk material the Bridgman, solid state recrystallization, and travelling heater methods have been developed. The most successful implementation of the bulk growth technique was the travelling heater method which allows to grow up to 5-cm diameter single crystals of high structural and electrical quality. Bulk crystal growth is currently used to support the first generation of photoconductive HgCdTe array fabrication.

The LPE is the most matured method for preparation of high-quality HgCdTe epitaxial films (Capper, 2011). Two different approaches have been developed: growth from Te and Hg solutions at 420-600^{0}C and 380-500 ^{0}C, respectively. Dipping, tipping and sliding boat techniques have been used to grow both thin and thick films. The tipping and dipping techniques have been implemented by using both tellurium- and mercury-rich solutions, but only tellurium-rich solutions have been used with the sliding boat. Both Hg- and Te-rich LPE can produce material of excellent compositional uniformity and crystalline quality (Capper, 2011).

MBE and MOVPE growth of HgCdTe is performed at much lower temperatures compare to LPE. Growth temperature is typically around 200-350 ^{0}C in MOVPE and around 150-200 ^{0}C in MBE (Garland, 2011; Maxey, 2011). Due to the weak Hg-Te bond, the evaporation of this material is not congruent. Consequently, Hg re-evaporates from the growth surface, leaving the surface Hg-deficient. Because of a large flux of Hg is necessary in this case, it means that the MBE growth temperature window is extremely small for HgCdTe and precise control of the growth temperature is highly desirable (Arias, 1994; Arias, 1994a). Furthermore, to obtain a desired cut-off wavelength within ±6%, one must control the Cd composition x within ±0.002 (Garland, 2011).

Routinely produced CdTe and CdZnTe (Zn~ 3-4%) single crystals are used as substrates for MBE and MOVPE growth. Today, the largest commercially available CdZnTe substrates are limited to approximately 7x7 cm^2 area. At this size, the wafers are unable to accommodate more than two 1024x1024 FPAs (Rogalski, 2009). Also, the quality of epitaxial layers is influenced by poor thermal conductivity, compositional nonuniformity, native defects, surface roughness and imperfect surface flatness CdZnTe substrates.

The most used dopants for MBE and MOVPE growth are arsenic as the p-type dopant and indium and iodine as the n-type dopants. There are no problems with in-situ incorporation of indium and iodine and low-temperature post-growth annealing is used to optimize the structural and electronic properties of the doped material, but is not required for activation of dopants. The MBE growth of HgCdTe is optimally performed under Te-rich conditions (Arias, 1994), thus Hg vacancies are the dominant native point defects in as-grown layers. Due to this reason the in-situ incorporated As occupy Hg sites in a lattice and annealing is required to its activate as a p-dopant by transferring to Te sites. The standard activation anneal is 10 min at ~425 ^{0}C followed by 24h at ~250 ^{0}C under a Hg overpressure to fill the Hg vacancies. The high-temperature annealing it limits many of the low-temperature-growth advantages of MBE. Therefore, attempts have been made to obtain the p-type activation of As either as-grown or after only a low-temperature anneal (Garland, 2011).

The use of alternative Si and GaAs substrates are attractive in IR FPA technology due to several reasons including available Si substrates with large sizes (up to 300 mm diameter), lower cost and compatibility with semiconductor processing equipment. The match of the coefficient of thermal expansion with Si readout circuit allows fabrication of very large FPAs exhibiting long-term thermal cycle stability. Of course, the large lattice mismatch between HgCdTe and alternative substrates (Si: 19%, GaAs: 14%) causes a high dislocation density (typically from mid-10^6 to mid-10^7 cm^{-2}) at the epilayer/substrate interface due to propagating of dislocations in the growth direction. The high dislocation density is detrimental to device performance because of the effect of dislocations on minority carrier lifetimes (Shin, 1992). For comparison, in epitaxial layers grown on CdTe and CdZnTe substrates the dislocation density is less than mid-10^5 cm^{-2}. The dislocation density below this value is believed to be not a serious problem unless they form clusters under device contacts. Despite these difficulties, epitaxial growth of high quality HgCdTe on 4-inch CdTe/Si substrates has been demonstrated for MWIR applications (Maranowski, 2001). Also, large area high quality HgCdTe epilayers were grown by MBE on 100 mm diameter (211)Si substrates with a CdTe/ZnTe buffer layer (Bornfreund, 2007). Epilayers of HgCdTe with extremely uniform composition and extremely low defects density were demonstrated by Peterson et al. (Peterson, 2006) on 4- and 6-inch diameter silicon substrates.

The lattice matched and alternative substrates with (100) and (211) orientations are commonly used. The best crystalline quality is obtained on the substrates with the slightly misoriented surfaces. For instance, the MOVPE growth on (100) substrates misoriented from the (100) plane by a few degrees is useful to suppress the formation of pyramid-shaped macrodefects, known as hillocks (Maxey, 2011). Today almost all growth is carried out on (211) substrates, which have (111) terraces with (100) steps (Garland and Sporken, 2011).

The MBE and MOVPE technology of HgCdTe has developed to rather high level at which epitaxial layers grown on bulk CdTe and CdZnTe substrates have characteristics comparable to those of LPE material.

3.1.2. Growth Of Insb And Inas

Due to the relatively low melting point of (525 ^{0}C) and small saturation vapor pressure, InSb bulk crystals can be grown using different growth techniques: Czochralski and horizontal Bridgman technique, travelling heater method and zone melting (Hulme and Mullin, 1962; Liang, 1966; Parker, 1965; Benz and Müller, 1975; Bagai, 1983). Technology of InSb bulk crystals are well matured. InSb is the most perfect material among the III-V semiconductors available to data. Typically 10-100 cm^{-2}etch pit density is specified in a commercially available InSb. However, the best result is only 1 etch pit in 50 cm^{-2}. In the ultra-high pure InSb bulk crystals the carrier concentration can be lower 10^{13} cm^{-3}.

In contrast to indium antimonide, InAs possesses an appreciable vapor pressure at the melting point (943 ^0C), because of the equilibrium vapor is constituted of the more volatile component (As) whose condensation and sublimation temperature lies below the melting point. However, even at this case, there is a strong tendency to form stoichiometric compound. InAs bulk crystals can be grown using liquid covering Czochralski or vertical gradient freeze method. Because of purification of InAs is more difficult than InSb, the residual electron concentration of InAs bulk crystals is about of $2 \cdot 10^{16}$ cm^{-3}.

Gettering effect of lead and rare earth elements (ytterbium and gadolinium) in LPE of InAs has been studied by several groups (Baranov, 1992, 1993; Voronina, 1999; Gao, 1999). The gettering effect of lead is attributed to the formation of stable insoluble aggregates composed of indium tellurides, selenides and sulphides. As a result, epitaxial films with the concentration of electrons $\sim^{15}$ cm^{-3} and mobility 9.1 10^4 cm^2/V s at 77 K have been grown. In these films reduction of the concentration of structural defects was also observed. The simultaneous introduction of controlled amounts of lead and rare earth elements makes it possible to prepare high resistivity films with the compensation degree 0.6-0.9. In the compensated films the electron concentration $3 \cdot 10^{15}$ cm^{-3} was achieved. The defect complexes of n-type were also observed in these films.

Various efforts have been made to adopt LPE growth of InSb (Kumagava, 1973; Mengailis, 1966;Holmes, 1980). However, there is a small number of reports on successful growth of InSb films by this method. Recently Dixit and co-authors reported growth of high-quality films on (001) semi-insulating GaAs substrate in a boat-slider type LPE unit (Dixit, 2002; Dixit, 2002a).

Heteroepitaxy of InSb and InAs has been achieved on Si and GaAs substrates with MBE and MOCVD technology (Razeghi, 2003). To overcome the lattice mismatch (>19%), the MBE of InSb on Si substrates was performed using CaF_2 and stacked BaF_2/CaF_2 buffer layers. The room temperature electron mobility of 65000 cm^2/V s (n $\approx$2 10^{16} cm^{-3}) was obtained in an 8 µm-thick film grown on a Si substrate with 0.3 µm CaF_2 buffer layer. The 77 K mobilities were at least an order of magnitude lower than the room temperature values. This behavior of electron mobility is attributed to electron scattering on dislocations arising from both lattice and thermal strains (Liu, 1997).

Epitaxial layers of InSb were grown directly on InSb, GaAs and GaAs-coated Si substrates with MBE and a low pressure MOCVD techniques (Razeghi, 2003). The quality of epitaxial films has been shown to depend critically on the growth conditions and preparation of substrates. In order to get high crystal quality InSb and GaAs substrates, directed 2^0 off the (100) toward (110) direction were used. The X-ray rocking curve FWHM, electron concentration and mobility was found to depend on the thickness of films due to influence of highly dislocated interface. In 3.6 μm thick InSb film the electron mobility was 56000 $cm^2/V{\cdot}s$ at 300 K and close to 80000 $cm^2/V{\cdot}s$ at 77 K. The background electron concentration at 77 K was of the order of 10^{16} cm^{-3}. Excellent uniformity (within the ±3 arcs variation of FWHM) was detected for a 10 μm thick InSb layer grown by MBE on a 3-inch semi-insulating GaAs substrate. The FWHM decreases with thickness as the dislocation density decreases due to the greater distance between the surface and the highly dislocated interface. The 300 K mobility close to that of bulk InSb (75000 $cm^2/V{\cdot}s$) was achieved in the films with thickness more than 2 μm. The temperature dependence of the electron mobility was peaked at 77 K and decreased at lower temperatures due to the dislocation scattering at the InSb/GaAs interface (Razeghi, 2003).

The MBE growth of InAs has been reported by several groups (Yano, 1997; Kalem, 1998). InAs epitaxial layers with thicknesses ranging from 0.5 up to 6.2 μm was grown on (100) oriented semi-insulating GaAs substrates. As in the case of InSb films, the properties of InAs films was influenced by the growth conditions and InAs/GaAs interface structure. The electron mobility at room temperature is 1.8 10^4 $cm^2/V{\cdot}s$ (n=6.1 10^{15} cm^{-3}) and peaks at about liquid-nitrogen temperature with a value of 5.173 10^4 $cm^2/V{\cdot}s$ (n=3.1 10^{15} cm^{-3}) for a InAs layer with thickness of 6.2 μm. It is shown that the temperature dependence as well as the magnitude of the mobility can be explained by a combined impurity-phonon-dislocation scattering mechanism. The dislocation densities of the order of 10^6 cm^{-2}were found. High-quality InAs epilayers on the GaAs substrates have been grown by MBE (Chen,2000; Cai, 2003). The growth was carried out as a two-step process: InAs layers were grown under As-rich conditions on InAs prelayers grown directly on the GaAs substrates under In-rich conditions. The optimized growth condition for this method from the Raman spectroscopy and the low-temperature photoluminescence was the following: first InAs is grown 20 nm thick under In-rich conditions at 500 ^{0}C with the appropriate V/III ratio of 8, then InAs is continuously grown under As-rich conditions at 500 ^{0}C with the appropriate V/III ratio of 10–23. Also, a two-step growth method consisting of a 400 ^{0}C prelayer followed by deposition of the thick bulk layer at higher growth temperatures has been reported by Watkins et al. (Watkins, 1995). High purity InAs epilayers were grown on GaAs substrates by MOCVD technique. Temperature dependent Hall measurements between 1.8 and 293 K showed a competition between bulk and surface conduction, with average Hall mobilities of $1.2 \cdot 10^5$ $cm^2/V \cdot s$ at 50 K. Large changes in the temperature dependent transport data are observed several hours after Hall contact formation and appear to be due to passivation of the surface accumulation layer by native oxide formation. The highest electron mobilities were observed in InAs films grown by MOCVD at reduced growth temperature (Partin, 1991). Electron mobilities as high as 21000 $cm^2/V \cdot s$ at 300 K were obtained for a film only 3.4 μm thick. From the depth dependence of transport properties it has been found that in the grown films electrons are accumulated near the air interface of the film, presumably by positive ions in the native oxide. The scattering from dislocations was greatly reduced in the surface accumulation layer due to screening by a high density of electrons. These dislocations arise from lattice mismatch and interface disorder at the film-substrate interface, preventing these films from obtaining mobility values of bulk indium arsenide. More or less successful attempts to grow quality InAs epitaxial films were also reported in numerous papers (Fukui, 1979;Haywood, 1990; Egan, 1995; Huang, 1995; von Eichel-Streiber, 1997; Watkins, 1997).

Technology of InSb bulk crystals is more mature than HgCdTe. Good quality substrates with more than 7 cm diameter are commercially available (Micklethwaite, 2000). However, technology of InSb and InAs epitaxial layers is not matured to the level suitable for device applications.

3.2. Junction Formation Techniques

3.2.1. Hgcdte Photodiodes

Different HgCdTe photodiode structures have been developed including mesa, planar and lateral homojunction and heterojunction structures. The history and current status of HgCdTe IR photodiodes has been reviewed in numerous monographs and review articles (Capper and Elliot, 2001;Norton,1999; Rogalski, 2011; Rogalski et al., 2000; Chu and Sher, 2010; Reine, 2000; Baker and Maxey, 2001; Sher, 1991). The junctions have been formed by numerous techniques including impurity diffusion, ion implantation, growth of doped epitaxial layers from vapor or liquid phase. In the early stages of device technology Hg in- and out-diffusion has been used for the junction formation in HgCdTe bulk material. At present time ion implantation is commonly used technique for formation n-on-p and p-on-n homo- and heterojunctions. The n^+-p homojunction can be formed by implantation of different ions into a vacancy doped p–type material, but B and Be are the most frequently used for this purpose. The doses of 10^{12}–10^{15} cm^{-2} and energies of 30–200 keV are used for the junction formation. First devices were fabricated on bulk p-HgCdTe single crystals with the hole concentration of the order of 10^{16} cm^{-3}. But in the early 1990s bulk crystals were replaced by LPE material. To-day, epitaxial films grown by LPE on CdZnTe lattice matched substrates are the best structural quality material, and they are successfully used in the homojunction technology. An ion implantation was also adapted to produce P^+-n heterojunctions (P means wider gap material). In both structures the lightly doped 'base' region with the doping concentration below 10^{16} cm^{-3} determines the dark current and photocurrent. Indium is most frequently used as a well-controlled dopant for n-type doping due to its high solubility and moderately high diffusion. Arsenic proved to be the most successful p-type dopant that are used for fabrication of stable junctions due to very low diffusivity. The important advantage of P+-n structure is that the thermal generation of carriers is effectively reduced in wider bandgap material.

The significant step in the development of HgCdTe photodiodes has been made by Arias with co-authors (Arias, 1993). They proposed the double-layer planar heterostructure (DLPH) photodiodes. The photodiodes were realized by incorporating a buried narrow-bandgap active layer in the DLPH configuration. The planar devices were formed using a $Hg_{1-y}Cd_yTe/Hg_{1-x}Cd_xTe$ ($y>x$) heterostructure grown by MBE on CdZnTe substrate. An important feature of the DLPH approach is a planar p-doped/n-doped device geometry that includes a wide-bandgap cap layer over a narrow-bandgap 'base' layer. The formation of planar photodiodes was achieved by selective implanting of arsenic through a ZnS mask followed by diffusing the arsenic (by annealing at high temperature) through the cap layer into the narrow gap base layer. After that the structures were annealed under Hg overpressure. The first high-temperature annealing was carried out to diffuse the arsenic into the base layer and to make the doped region p-type by substitution of arsenic atoms on the Te sub-lattice, while the second low-temperature one was carried out to annihilate Hg vacancies formed in the HgCdTe lattice during high-temperature process. In DLPH photodiodes significant reduction in tunneling current and surface generation-recombination current has been achieved. The architecture of mesa and planar heterostructure photodiodes are shown in Fig.1. The back-illuminated heterostructure photodiodes prepared from both LPE and MBE material grown on CdZnTe substrates have the highest performance achieved to-day (Arias, 1993; Bajaj, 2000; Rogalski, 2000).

The serious disadvantage of CdZnTe substrates is the thermal expansion coefficient mismatch with Si used for the read-out integrated circuit. To overcome this problem, growth of HgCdTe on alternate substrates such as sapphire, Si and GaAs has been developed. LPE, MBE and MOCVD techniques were used for this purpose. However, these alternate substrate materials suffer from a large lattice mismatch with HgCdTe, leading to a higher defect density in the HgCdTe material and consequently reducing the detector performance (Bajaj, 2000). Despite these disadvantages the large, 1024-1024 and 2048-2048, HgCdTe FPAs operating in SWIR and MWIR spectral bands were grown on alternative substrates (Kozlowski, 1999; Bajaj, 2000; Golding, 2003; Tribolet, 2003). Their performance is comparable to performance of FPAs prepared on lattice matched bulk substrates, with the same spectral cut-off.

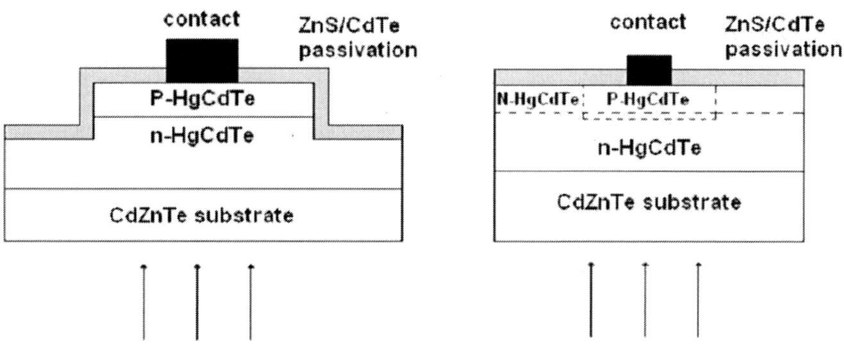

Figure 1. Schematic cross section of mesa (left) and planar (right) HgCdTe IR photodiodes. The photodiodes are illuminated through the wide bandgap substrate.

While ion implantation of As requires activation of the implanted atoms at relatively high temperatures, an alternative technology (ion milling or plasma induced type conversion) has received considerable attention during the past few years. Currently plasma induced type conversion in HgCdTe is regarded as an alternative to ion implantation junction formation technology (Agnihotri, 2002). The post-implant annealing is not needed in this technology. Reactive ion etching (RIE) induced type conversion and junction formation have been observed in a vacancy doped p-HgCdTe using H_2/CH_4plasma. The junction depth could be adjusted from 2 to 20 μm.

3.2.2. Insb and Inas Photodiodes

High performance InSb detectors have been fabricated with bulk material for decades. Typically, the p-on-n junctions are prepared on bulk crystals of n-type conductivity with electron concentration in the range 10^{14} - 10^{15} cm^{-3} and mobility of the order of (2-6) 10^5 cm^2 V^{-1} s^{-1} at T = 77 K. Be ion implantation and thermal diffusion of Zn and Cd seems to be the most frequently used technological methods which allow to obtain sharp p^+-n junctions (Mozzi and Lavine, 1970; Hurwitz and Donnelly, 1975; Rosbeck, 1981; Nishitani, 1983; Fujisada, 1985). The current status of InSb photodiode technology have presented by Wimmers et al. (Wimmers, 1983; Wimmers, 1988). Current manufacturing device processes require that bulk materials should be thinned before or after the junction preparation. Using this technique, hybrid FPAs were produced (Fowler, 1996; Hoffman, 1991; Parrish 1991). An array size of 1024-1024 was possible because the InSb detector material was thinned to <10 μm (after surface passivation and hybridization to a readout chip) which allows it to accommodate the InSb/silicon thermal mismatch. Certainly, the substrate thinning process is a very delicate process that can lower the yield and the reproducibility of the photodiodes, but it is used commercially because of LPE of InSb is not developed. Several attempts have been made to manufacture InSb photodiodes by LPE (Kosogov and Perevyaskin, 1970; Kazaki, 1976).

Currently the major efforts are focused on photodiodes grown by MBE and MOCVD methods. An overview *of* these technologies has been done by Razeghi (Razeghi, 2003). The InSb photodiodes were grown on 3-inch Si and (111)GaAs substrates. The InSb photodiodes typically consisted of a 2 μm n^+ region (~10^{18} cm^{-3} at 77 K), a ~6 μm unintentionally doped region (n = 10^{15} cm^{-3} at 77 K) and a ~0.5 μm p^+ (~10^{18} cm^{-3}) contact layer. The crystallinity of these structure was excellent as confirmed by the X-ray diffraction, which showed FWHM <100 arcs for structures grown on (111)GaAs and Si substrates. Photodiodes were fabricated with 400 × 400 $μm^2$ mesa structures by photolithography and wet chemical etching. These devices showed excellent response of about 1000 V/W, which is comparable to that of bulk detectors, with detectivities of ~3 10^{10} cmHz$^{1/2}$/W at 77 K. It was shown that photodiodes can operate up to room temperature, even though they were not optimized for this purpose.

A miniaturized InSb photovoltaic infrared sensor that operates at room temperature was developed byKuze with co-authors (Kuze, 2007). The InSb sensor consists of an InSb p^+-p^--n^+ structure grown on semi-insulating GaAs (100) substrate, with a p^+-$Al_xIn_{1-x}Sb$ barrier layer between p^+ and p^-layers to reduce diffusion of photoexcited electrons. The optimum Al composition and thickness of AlInSb barrier layer was found to be x=0.17 and 20 nm, respectively. Typical responsivity of 1.9kV/W, output noise of 0.15 μV/Hz$^{1/2}$ and detectivity of 2.8 10^8 cm Hz$^{1/2}$/W was measured at 300K. InSb high-speed photodetectors were grown on semi-insulating GaAs substrate with 0.1 μm GaSb buffer layer using MBE (Kimukin, 2003). After the buffer layer a 1.5 μm thick n-InSb layer, 1.5 μm thick n-InSb layer, and finally 0.5 μm thick p- InSb layer were grown. The n-active layer was unintentionally doped to 2-3 10^{15} cm^{-3}. Tellurium and beryllium were used as the n- and p-layer dopants, respectively. Doping level was 10^{18} cm^{-3} for both highly doped layers to decrease the serial resistance. The developed photodetectors can operate at room temperature. The responsivity 1.3 A/W was measured at 1.55 μm wavelength at room temperature. InSb photodiodes grown by MBE on GaAs coated Si substrates have been reported (Besikci, 2000; Ozer and Besikci, 2003). The peak detectivity of ~$1 \cdot 10^{10}$ cm Hz$^{1/2}$ W^{-1}at 80 K has been measured under backside illumination without anti-reflection coating. Differential resistance at 80 K is shown to be limited by ohmic leakage under small reverse bias and trap assisted tunneling under moderately large reverse bias.

InAs photodiodes are mainly manufactured by ion implantation and diffusion methods (McNally, 1970; Astahov, 1992). The p-n junctions prepared by Cd diffusion into n-InAs single crystals were investigated by Tetyorkin et al. (Tetyorkin, 2011). The photodiodes grown by MBE on InAs substrates have been reported (Kuan, 1996; Lin, 1997). The fabrication of InAs photodiodes on GaAs and GaAs-coated Si substrates by MBE have also been reported (Dobbelaere, 1992).

In conclusion, during the past ten years the impressive progress has been made in the growth of narrow-gap semiconductors by MBE, MOVPE and MOCVD on silicon, GaAs and sapphire substrates. It is expected that in the near future HgCdTe and InSb IR photodiodes will be grown directly on silicon for the low cost detectors with bi-spectral and multi-spectral capability.

3.3. Thermal Annealing

In HgCdTe photodiodes prepared on bulk crystals and LPE films the active absorption region should be p-type with the hole concentration of the order of 10^{16} cm^{-3}. Because of the as-grown materials are strongly p-type with vacancy concentration $p>10^{17}$ cm^{-3}, they should be annealed. Due to importance of annealing, it has been intensively studied in HgCdTe (Capper, 1994; Capper, 1997; Capper 2011). The most important parameters of annealing are temperature, Hg vapor pressure and annealing duration. In HgCdTe (x = 0.2) Vydyanath determined the free hole concentration at 77 K as a function of the Hg partial pressure, P_{Hg}, at anneal temperatures between 140 and 655 ^{0}C. The concentration of doubly-ionized vacancies $[V_{Hg}]$ is given by (Vydyanath, 1991):

$$[V_{Hg}]P_{Hg} = 1.7E28 \exp(-1.67\ eV/kT) cm^{-3} atm$$

The hole concentration at 77 K is assumed to be equal to the concentration of vacancies. It was proved that there is no essential difference in the annealing behavior of bulk crystals and LPE epilayers. As seen, the same vacancy concentration can be obtained by combination of temperature and Hg vapor pressure. Obviously, samples annealed at lower temperature and pressure conditions would also have the lower concentration of interstitials and potentially the higher minority carrier lifetime. However, to reduce density of Te precipitates and dislocations high-temperature annealing should be performed (Capper, 2011). In epilayers grown on lattice-matched substrates (CdTe, CdZnTe) the dislocation density is in low-10^4 to mid-10^5 cm^{-2}. At the same time, in epilayers grown on alternative substrates the dislocation density exceeds 10^6 cm^{-2}. In LWIR epilayers grown by MBE on (211) GaAs and Si substrates the dislocation densities as low as 2.3 10^5 cm^{-2} have been obtained using high-temperature cycling between 300 and 490 °C (Shin, 1992). It has been shown that epilayers grown by MOCVD require a higher thermal annealing temperature than MBE material and the difference in dislocation reduction between MBE and MOCVD HgCdTe materials is caused by dislocation movement under high-temperature and thermal stress conditions. A strong correlation between minority carrier lifetime and dislocation density was observed. The effect of temperature cycling on the dislocation density was also investigated by Farrell et al. (Farrell, 2011). In-situ and ex-situ thermal cycle annealing methods have been used to decrease dislocation density in CdTe and HgCdTe. During the MBE growth of the CdTe buffer layer, the growth was interrupted and the layer was subjected to an annealing cycle within the growth chamber under tellurium overpressure. During the annealing cycle the temperature is raised to beyond the growth temperature (290 → 550 °C) and then allowed to cool before resuming growth again. This process was repeated several times during the growth. The in-situ thermal cycle annealing resulted in almost a two order of magnitude reduction in the dislocation density. The decrease of dislocation density was attributed to the movement of the dislocations during the annealing cycles and their subsequent interaction and annihilation. To decrease the dislocation density in HgCdTe layers grown on CdTe/Si composite substrates, ex-situ annealing has been performed in a sealed quartz ampoule under a mercury overpressure. It was found that the primary parameters that affect dislocation density reduction are the annealing temperature and the number of annealing cycles.

At last, high temperature anneal in Hg vapor is used to activate the dopant by substituting arsenic atoms on the Te sublattice. A lower temperature anneal (200–250 ^{0}C), performed immediately after the high temperature anneal, annihilates the Hg vacancies formed in the HgCdTe lattice during high temperature treatment.

Bulk, LPE, MOCVD and MBE grown high purity material subjected to low-temperature (<300 ^{0}C) Hg reach annealing are converted to n-type with the electron concentration which depends on the doping level of residual donors. In HgCdTe material for LWIR photodiodes the minimum electron concentration that can be obtained is ~(2-10) 10^{14} cm^{-3} and the mobility is ~6 10^4 cm^2/V s (Chu and Sher, 2008).

Techniques for improving the control of the Hg vacancy concentration was also reported by Yang et al. (1985), who deduced experimentally a relationship between the concentration of Hg vacancies, the annealing temperature, and the temperature of a Hg source. They also derived theoretically an analytic expression for this relationship. The annealing for MBE grown material can be conducted in-situ under ultra-vacuum conditions. In-situ annealing is very useful for production purposes. The improved ex-situ and in-situ annealing processes were developed for MOCVD grown films (Madejczyk, 2005).

The performance of LWIR photodiodes can be improved by post-implant annealing (Bubulac, 1998). The dramatic decrease of the dark current in LWIR photodiodes was observed due to low temperature annealing at 120-150 °C (Ajisawa and Oda, 1995). The improvements was explained by changes in both carrier concentration profile and p-n junction position determined by interaction of interstitial Hg atoms with vacancies in the vicinity of the junction during the annealing process.

Very scarce data are available in the literature concerning the annealing of InSb and InAs, both materials and devices. The effect of rapid thermal annealing and sulfur passivation on the quality of reactively sputtered SiO_2 on InAs were investigated. Results show that both rapid thermal processing and sulfur passivation cause a reduction in leakage current and oxide fixed charges. Annealing at temperature higher than 400 ^{0}C caused degradation. Also, passivation started to loose its effectiveness when structures are annealed at 500 °C (Eftekhari, 1997).

4. CARRIER TRANSPORT AND RECOMBINATION MECHANISMS

4.1. Tunnelling Current In Hgcdte, Inas And Insb Photodiodes

Tunneling current was observed by many authors in IR photodiodes made of narrow-gap A_2B_6 and A_3B_5 semiconductors (Rogalski, 1995). However, its nature seems to be understood in rare cases. For instance, the trap-assisted tunneling (TAT) via single level in the gap introduced by point defects was proved to be the main reason for the excess current in HgCdTe IR photodides at rather small reverse biases followed by the direct band-to-band (BTB) tunneling current at higher biases (Nemirovsky, 1992; Rosenfeld, 1992; He, 1996). In these photodiodes the trap-assisted tunneling current is shown to be a source of the low-frequency 1/f noise. Similar results were also obtained in HgCdTe MIS structures (He, 1996). Dislocations are also known for a long time as a source of an excess current in semiconductor devices, especially when they intersect the depletion region of the p-n junction (Matare, 1971; Holt, 2007; Shikin, 1996; Whelan, 1969). As to InAs and InSb photodiodes the role of dislocations is not established clearly. Therefore, identification of the type of defects participating in the carrier transport in InAs and InSb IR photodiodes is a key problem for improvement of their performance. Usually TAT and BTB tunneling currents were analyzed in the reverse biased IR photodiodes. This analysis performed for the forward biased InAs and InSb photodiodes revealed new aspects of tunneling transport of carriers.

4.2. Dislocation-Assisted Tunnelling Current In The Forward-Biased Inas And Insb Photodiodes

The photodiodes were prepared on single-crystal substrates of n-type conductivity. In order to investigate effect of dislocations on the dark current, the substrates were cut from different parts of ingots grown by Bridgman technique. The density of dislocations in InAs substrates measured by the etch-pit method was of the order of 10^4 cm^{-2}. The damaged surface layers were removed using dynamic chemical-mechanical polishing in solution of methanol with 2% of Br_2. The electron concentration and mobility in the initial substrates were (2-3) 10^{16} cm^{-3} and (2-2.5) 10^4 $cm^2/V{\cdot}s$, respectively. The dislocation density was ranged from (1-2) 10^4 cm^{-2} in the central part of ingots up to $4 \cdot 10^5$ cm^{-2} at the periphery one. The p-n junctions were prepared by thermal diffusion of preliminary synthesized $CdAs_2$ into substrates in sealed evacuated quartz ampules. The mesa structures with active area A = 7 10^{-3} cm^2 were delineated using standard photolithographic technique. Then Zn and In contact pads were thermally deposited onto p- and n-type sides of the junction, respectively, followed by a heat treatment in purified hydrogen atmosphere. InSb photodiodes were manufactured by implantation of beryllium into appropriately prepared substrates followed by thermal annealing. In n-InSb substrates the electron concentration and mobility were of the order of (1-2) 10^{15} cm^{-3} (6-7) 105 $cm^2/V{\cdot}s$ at 77 K, respectively. The dislocation density in InSb substrates was not exceeded 5·102 cm-2. Photodiodes had planar structure with the junction area A = 1.33 10^{-2} cm^2.

The current-voltage characteristics are shown in Fig.2. As seen, the characteristics consist of two exponential parts. At lower bias voltages the forward current in InAs photodiodes increases with increasing the density of dislocations, whereas at higher voltages it has approximately the same magnitude for both photodiodes. Fig. 3 shows the temperature dependence of the forward current in InAs and InSb photodiodes measured at the bias voltage 10 mV. At low temperatures $T < 130$ K the current is weakly dependent on temperature. At the same time at higher temperatures it exhibits an activation character. To clarify the observed peculiarities the current-voltage characteristics were investigated in InAs photodiode subjected to ultrasonic treatment with frequency 5-7 MHz and intensity ~0.4 W/cm^2 during four hours at room temperature, Fig. 4.

The dark current as a function of bias voltage was measured immediately after ultrasonic treatment as well as after approximately one-year storage of photodiodes at laboratory condition. It must be pointed out that the ultrasonic treatment results in pronounced increase of the dark current at lower bias voltages, whereas those parts of the current-voltage characteristics measured at higher voltages were remained almost unchanged. Also, it is important to note that after storage of samples within one year the excess current caused by ultrasonic treatment is decreased to approximately the starting values.

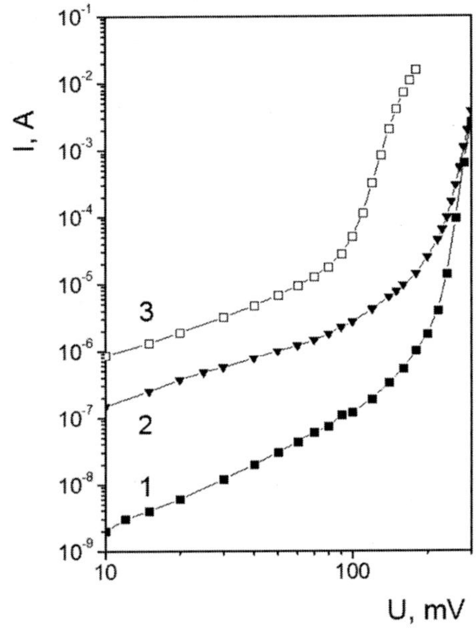

Figure 2. Current-voltage characteristics in InAs (1,2) and InSb (3) photodiodes at 77 K. Curves (1) and (2) refer to the dislocation density in substrates 4 10^4 cm^{-2} (1) and 2 10^5 cm^{-2}, respectively.

An explanation of experimental results is based on the assumption that dislocations intersecting the depletion region are responsible for the excess current at small forward biases. A model for tunneling current via dislocations intersecting the depletion region of the junction has been proposed by Evstropov et al. (Evstropov, 1997, 2000). Experimentally it has been also investigated by Ageev with co-authors (Ageev, 2009). According to this model, mobile carriers (holes and electrons) are moved along acceptor-like and donor-like dislocation lines which has been modeled by a chain of parabolic potential barriers with variable height.

In the symmetric junction the forward current flows due to direct recombination of electrons and holes at the middle of the depletion region. The current-voltage characteristics can be described by the formula

$$I = I_{01}\exp\left(\frac{e(U - IR_s)}{E_0}\right) + I_{02}\exp\left(\frac{e(U - IR_s)}{\beta kT}\right) \quad (1)$$

where I_{01} and I_{02} are the pre-exponential factors, E_0 is the characteristic energy; β is the ideality factor and R_S is the series resistance.

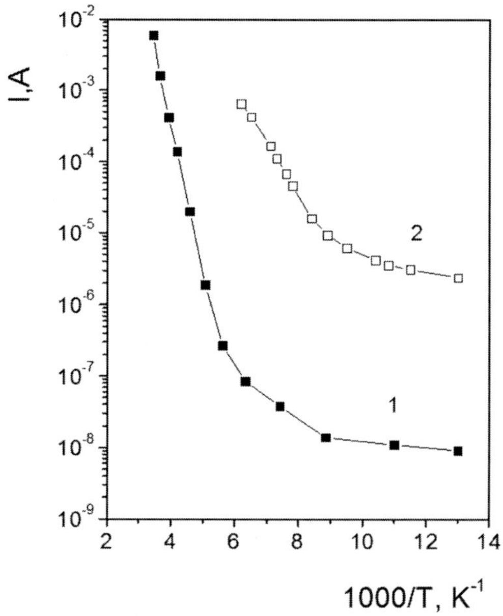

Figure 3. Temperature dependences of the forward current measured at 10 mV in InAs (1) and InSb (2) photodiodes

The temperature dependence of the pre-exponential factor in equation (1) is then given by:

$$I_{01} = e\rho v_D A\exp\left(\frac{eU_D}{E_0}\right) \quad (2)$$

where ρ is the density of dislocations, ν_D is the Debye frequency, U_D is the diffusion potential. The lifetime of carriers in the depletion region τ0 was determined from the relation $I_{02} = en_iWA/\tau_0$, where n_i is the intrinsic concentration of carriers, W is the depletion region width, A is the junction area.

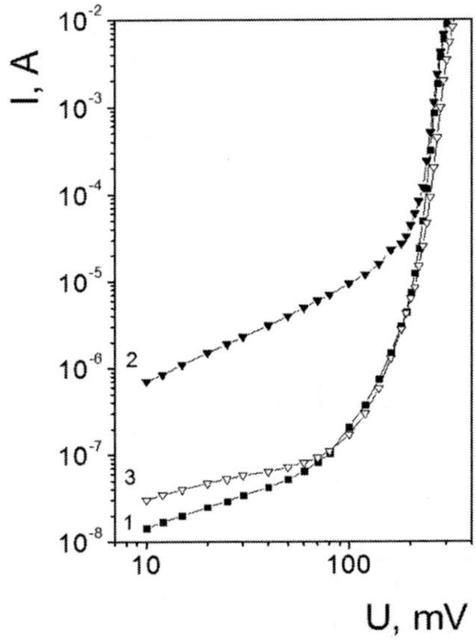

Figure 4. Current-voltage characteristics in InAs photodiodes before and after ultrasonic treatment (curves 1 and 2, respectively), and after one-year storage (3).

Because of in the investigated photodiodes the diffusion potential linearly depends on temperature (Sukach, 2005), the exponential dependence of I_{01} on temperature should be observed. Also, in accordance with Evstropov, the characteristic energy E0 is independent on the concentration of free carriers. These consequences of the analyzed model may be used for discrimination of the tunneling current via dislocations. For instance, by using typical experimental data for InSb photodiodes at 77 K ($I_0/A = 8.85 \cdot 10^{-5} A/sm^2$, U_D = 160 mV, E_0 = 29 meV and $\nu_D = 3.3 \cdot 10^{12}$ s^{-1}, which was determined from the known value of Debye temperature T_D = 160 K (Madelung, 1996), the dislocation density $\rho = 4.2 \cdot 10^4$ cm^{-2} was estimated. This value is almost two orders of magnitude higher than in the starting substrates. Relatively high density of dislocations can be explained by the fact that during the heat treatment the edge of the junction was not removed from the zone of radiation defects formed by ion implantation of beryllium in InSb. However, the same discrepancy between experimental and theoretical data was also observed in InAs photodiodes prepared by diffusion technique. Moreover, in the investigated InAs photodiodes the characteristic energy E_0 was found to be varied from ~30 meV up to ~60 meV in contrast to theoretical predictions. It must be stressed that values of E_0experimentally obtained in this study are close to those observed in diodes made of wide-gap GaP and SiC (Evstropov, 1997, 2000; Ageev, 2009). This means that the tunneling current via dislocations is characterized by the same features independent on semiconductor materials used for manufacture of diode structures.

The observed discrepancy between theoretical and experimental data may be caused by several reasons. First of all, in the model developed by Evstropov dislocation lines are assumed to be fully occupied by carriers and have a length of the order of the depletion region width. This seems to be not typical for dislocations in semiconductors (Matare, 1971; Holt and Yacobi, 2007; Shikin, 1995). Also, the presence of jogs, inclusions of impurity atoms, kinks, etc., results in a loss of translation symmetry along the dislocation line and spatial localization of mobile carriers. Thus only short dislocation segments can contribute to the direct current conduction (Holt, 2007; Kveder, 1985; Nitccki, 1985). It is supposed that this is the main reason why the direct current conduction along dislocation cores is not clearly demonstrated so far (Holt and Yacobi, 2007). Further, as originally proposed by Shockley (Holt and Yacobi, 2007) and according to Labusch (Labusch, 1982) and Labusch and Schröter (Labusch and Schröter, 1983) dislocations in semiconductors introduce one-dimensional energy bands into the gap, located near the conduction and valence band edges. Direct recombination transitions between these bands seem to be not effective. The much more effective is the recombination through deep defect states in the gap (Kveder, 2001; Seibt, 2009).

Further analysis of experimental data is based on assumption that the p-n junctions in the investigated diodes are non-homogeneous and there are two conduction paths for mobile carriers in the junction. The tunneling current flows via the dislocations intersecting the depletion region whereas the recombination current flows via homogeneous region free of dislocations. At low bias voltages the tunneling current is dominant, so the forward I-U characteristic is described by the first term inequation (1). Thus, the weak dependence of the forward current on temperature in Fig.3 can be qualitatively understood. With increasing the bias voltage this current is masked by the recombination current which can be explained within the well known SRH model (Sze, 1981). This change in the current mechanism is described by the second term in equation (1).

Experimental evidences exist that the recombination rate of minority carriers at dislocations in silicon depends strongly on dislocation decoration by transition metal impurities (Seibt, 2009). Due to the fact that the recombination of carriers captured at dislocation bands can be substantially enhanced by the presence of small amount of impurity atoms at the dislocation core, it is assumed that the low-temperature transport mechanism consists of several steps, namely: a) injection of electrons into the depletion region under the forward bias, b) capture of electrons on the dislocation core by tunneling transitions, c) electron transport along the undisturbed segments of the dislocation core and d) recombination of electrons with holes through the states in the gap related with 'native' core defects (such as jogs and kinks) or impurity atoms segregated to the dislocation core. In the case of the dislocation core can exchange electrons directly with the conduction band the energy E_0 is the dislocation barrier height. Using experimental values for the density of dislocations (~10^4 cm^{-2}) and the forward current (10^{-8}-10^{-7} A) it is easy to show that dislocations form equipotential lines. If we take into account that the dislocation resistivity is of the order of 10^{10} Ω/cm (Labusch, 1982), the voltage drop along a segment of dislocation of 10^{-5} cm is less than 10^{-4} V. So, it is likely that at low temperatures tunneling transitions of electrons to the dislocation core is the bottle-neck for the forward current in the investigated photodiodes. Also, it is possible that these transitions can occur via local states in the gap, associated with point defects or their precipitates surrounding dislocations. In this two-step process physical meaning of E_0 should be corrected taking into account the energy of these states. Because of in this study the pre-threshold intensity of ultrasonic treatment was used, experimental results can be explained by rearrangement of existing defects rather than generation of new point defects. In accordance with the vibrating string model of Granato-Luecke, the intensive sonic-dislocation interaction results in an effective transformation of the absorbed ultrasonic energy into the internal vibration states of a semiconductor stimulating different defect reactions (Granato and Luecke, 1956). The driving force of the long-term relaxation of the forward current may by stress and electric fields around dislocations.

In conclusion, the excess current experimentally observed in InAs and InSb photodiodes at forward biases is related to dislocations intersecting the depletion region. Pronounced effect of ultrasonic treatment on the forward current is explained by transformation of defects segregated around dislocations. A model for the carrier transport via dislocations is proposed.

4.3. Trap-Assisted Tunnelling Current In The Reverse-Biased Inas And Insb Photodiodes

The effect of traps in the depletion region of a photodiode on the TAT current was considered by several authors (Wang, 1980; Kinch, 1981; Nemirovsky, 1989, 1991; Rosenfeld and Bahir, 1992; He and Celik-Butler, 1995). The calculation of the TAT current in a reverse-biased photodiode is carried out using several simplifying assumptions: the p-n junction is abrupt with a linear variation of potential (constant electric field) across the depletion region; the traps are uniformly distributed; the initial states are occupied whereas the final states are empty. Under these assumptions, the TAT current is proportional to the trap density, but depends exponentially on the trap-ionization energy and the electric field strength. Within this model the observed soft reverse breakdown current-voltage characteristics were adequately explained.

For instance, in Fig. 5 and 6 are shown typical current-voltage characteristics and 1/f noise spectra measured in n^+-p HgCdTe (x=0.22) photodiodes. The photodiodes were prepared by boron implanting into epitaxial films grown by LPE method, followed by surface passivation and low-temperature post-implanting anneal. In the calculation of TAT current thermal and tunnel transitions from the valence band to deep defect states in the gap followed by tunnel transitions to the conduction band were taken into account. That is, the TAT current is given by

$$J_{tat} = qWN_t\left(\frac{1}{\omega_v N_v + c_p p_1} + \frac{1}{\omega_c N_c + c_n n_1}\right)^{-1} \tag{3}$$

were $\omega_c N_c$ and $\omega_v N_v$ are the tunneling rates. In order to fit experimental and calculated data it has been supposed that are non-uniformly distributed through the depletion region (Ivasiv, 1999). The best fit was obtained for the acceptor-like traps with energy $E_t = 0.72E_g$ above the top of the valence band and the capture rates for holes and electrons $C_p \approx 10^{-7} - 10^{-6}$ cm^3/s and $C_n = (0.1\text{-}0.01)C_p$, respectively. The determined trap energy correlates well with the previously used $E_t=0.75E_g$ in photodiodes prepared by boron implantation to bulk material (Nemirovsky, 1991). Note that in photodiodes investigated by

Nemirovsky et al., the TAT current was dominant by donor-like traps with the capture rates C_p=(10^{-10}-10^{-9}cm^3/s) and C_n = (10-100)C_p. Our data are in accordance with the study performed by Rosenfeld and Bahir for acceptor-like centers in HgCdTe photodiodes (Rosenfeld and Bahir, 1992).

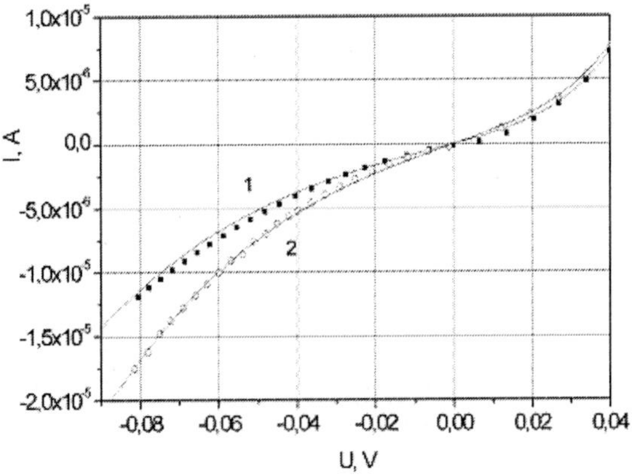

Figure 5. Measured (dots) and calculated (solid lines) current-voltage characteristics of n$^+$-p photodiodes at 77 K. Parameters of traps: E_t=0.72E_g, C_p=1 10^{-7} cm^3/s, C_n=0.01C_p, N_t=3.2 10^{15} and 4.8 10^{15} cm^{-3} for curves 1 and 2, respectively.

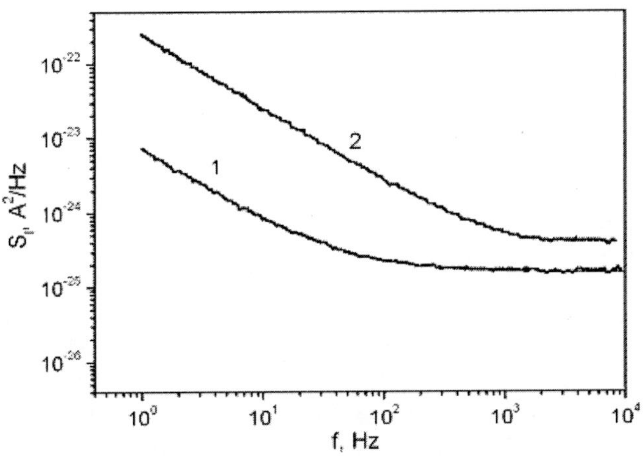

Figure 6. Noise spectra for the photodiodes with R_0A=1.0 Ω cm^2 (1) and R_0A=0.3 Ω cm^2 (2) at 77 K.

The correlation was found between the TAT current at small reverse bias voltage U ≤ 0.1 V and 1/f noise. In the case of the total dark current was completely determined by TAT mechanism the 1/f noise was observed up to frequencies 10^4 Hz. Taking this into account the noise current was calculated by the formula

$$I_n = \alpha \left(\frac{I_{TAT}}{\sqrt{f}} \right) \tag{4}$$

From the fitting calculation the values of the constant $\alpha = 10^{-8} - 10^{-7}$ were found. Earlier similar correlation was observed by Nemirovsky et al.. (Nemirovsky, 1989; Nemirovsky, 1992).

It seems that parameters of traps determined from the fitting calculations within this model may be used for rough estimations only. More accurate calculations of the TAT current has been performed byKrishnamurthy et al. (Krishnamurthy, 2006). The TAT current has been calculated for a linearly varying electric field in the depletion region of the p-n junction and self consistently obtained trap-occupation probability. The calculations showed that the reverse-bias dark current changes considerably both in magnitude and in shape. For a better interpretation of the observed dark currents and an estimation of trap density, these improvements should be taken into account.

In order to explain experimental data in InAs photodiodes it has been assumed that the tunneling current is controlled by small areas of the junction which are characterized by large deviation of impurity concentration from the mean value (Sukach, 2005). For instance, non-uniform distribution of impurity atoms can be realized around dislocations (Cottrell atmosphere) or at the periphery of the junction. This results in increase of electric field in the junction over the value given by the equation (1) for more or less uniform distribution of charged defects in the junction. We also assumed that the tunneling current in these areas is caused by the trap-assisted tunneling described with the modified values of the junction area and electric field strength in the junction. For this purpose, the dislocation was modeled by the effective area $A_{eff} \approx 1\ \mu m^2$ with increased concentration of charged defects. Their concentration was determined from the fitting calculation of I-U curves. The density of dislocation was assumed to be of the order of 10^4 cm^{-2}. The electric field around a dislocation may be determined from the Poisson equation. However, as the first approximation there has been assumed that the electric field around dislocations may be estimated using formulas for the abrupt p-n junction (Sze, 1981). Because of the tunneling rates $\omega_c N_c$ and $\omega_v N_v$ are exponentially depend on the electric field strength, the TAT current through these regions are exponentially large in comparison with the uniform regions of the junction. The trap-assisted tunneling current was calculated for the following cases: i) traps are exchanged with both bands by thermal and tunnel transitions of carriers (curve 1), ii) tunnel transitions of carriers from the valence band to traps followed by thermal and tunnel transitions to the conduction band (curve 2), tunnel transitions of carriers from the valence band to the conduction band through traps. The best fit was obtained for the energy of traps $E_t = E_g/2$ and their concentration in the range from $\sim 10^{13}$ to $\sim 10^{14}$ cm^{-3}. These values seem to be reasonable for InAs. The concentration of charged defects determined from the fit of the calculated and measured data was found to exceed $4 \cdot 10^{16}$ cm^{-3}. This value is more than one order of magnitude higher than the mean value of the free carriers concentration determined from the capacitance-voltage measurements.

Some arguments in favor of the measured current in InAs photodiodes is related to Cottrell's atmospheres around dislocations have been obtained from investigations of effect of ultrasonic treatment on the current-voltage characteristics (Sukach and Tetyorkin, 2009). In the photodiodes subjected to ultrasonic vibration with frequency 5-7 MHz and pre-threshold intensity ~0.4 W/cm^{2}prononced increase of the reverse current was obseved. The current is relaxed down to the starting value during nine-month storage of photodiodes at laboratory conditions. Experimental results are explained by transformation of existing complex defects rather than generation of new point defects. Most probably that this transformation is connected with Cottrell's atmospheres around dislocations which intersect the p-n junction. In accordance with the vibrating string model of Granato-Luecke (Granato and Luecke, 1966), the intensive sonic-dislocation interaction results in an effective transformation of the absorbed ultrasonic energy into the internal vibration states of a semiconductor stimulating different defect reactions. The driving force of the observed relaxation may be deformation and electric fields around dislocations.

4.4. Recombination Mechanisms

The carrier lifetime in HgCdTe, InSb and InAs narrow-gap semiconductors is determined by three principal recombination mechanisms: radiative, Auger and SRH. The first two mechanisms are intrinsic, whereas SRH recombination is not intrinsic because it is carried out with assistant of deep defect states in the gap. In principle, SRH recombination can be suppressed by reducing the concentration of recombination centers. Ten types of possible band-to-band Auger recombination processes in n- and p-type semiconductors were determined by Beattie (1962). The Auger 1 recombination mechanism in n-type material with InSb-like parabolic band structure was firstly considered by Beattie and Landsberg (Beattie and Landsberg, 1959). The Auger 7 process is important in p-type material (Beattie and Smith, 1967; Petersen,1970; Takeshima, 1972; Casselman and Petersen, 1980; Casselman, 1981). For the nonparabolic band structure, the $|F_1F_2|$ dependence on k, and nongenerate statistics appropriate expressions for the Auger 7 recombination process has been deduced by Beattie and Smith (1967).

The well known problem in the Auger recombination processes is the uncertainty in the carrier lifetime introduced by the overlap integrals F_1 and F_2 of the periodic part of the electron wave functions. As was shown by Petersen (1970, 1981), the dependence of the product $|F_1F_2|$ on the wave vector k should be taken into account in p-type materials. However, in practice the constant value of $|F_1F_2|$ in the range 0.1-0.3 is used for calculations of the carrier lifetime (Rogalski, 1995). This results in scatter of the calculated data within an order of magnitude. The detailed analysis of recombination process in HgCdTe can be found in numerous review articles and books (Beattie and Landsberg, (1959);Petersen, 1981; Capper,1994; Rogalski, 2011; Chu and Sher, 2010).

Due to Capper (1994), in n-type HgCdTe for low values of composition ($x<0.25$) and carrier concentrations $>10^{15}$ cm^{-3} the Auger 1 recombination process is dominant, particularly at high temperatures (>100 K). The SRH recombination is important at lower values of carrier concentration and at low temperatures. In LPE material grown from Hg-rich solution higher values of lifetime than in corresponding material grown by other techniques were observed, presumably due to a lower level of recombination centers related with Hg vacancies. Dislocations can also contribute to the SRH recombination when present at high densities. The measured data in p-type HgCdTe indicated that the Auger 7 recombination does not limit the carrier lifetime. It is believed that the SRH recombination can explain most of the experimental data (Fastow, 1990).

The Auger recombination process in narrow-gap semiconductors with the three- and four-band Kane models of band structure was reexamined by Gelmomnt et al. (Gelmont, 1978; Gelmont,

1981;Gelmont, 1982; Gelmont and Sokolova, 1982). The appropriate calculations of the carrier lifetime in InAs based on Gelmont's theory has been performed by Tetyorkin and co-authors (Tetyorkin, 2011). The calculated dependences of the lifetime as a function of the carrier concentration in InSb is shown in Fig. 7 and 8.

The generation rate for the Auger 1 process g_{A1} obtained by Beattie and Landsberg is given by

$$g_{A1}=\frac{8(2\pi)^{5/2}e^{4}m_e^{*}\,|F_1F_2|^{2}n_o}{h^{3}\varepsilon^{2}(1+\mu)^{1/2}(1+2\mu)}\left(\frac{kT}{E_g}\right)^{3/2}\exp\left[-\left(\frac{1+2\mu}{1+\mu}\right)\frac{E_g}{kT}\right] \tag{5}$$

According to Gelmont, the generation rate for the Auger 7 is

$$g_{A7}^{G}=\frac{18m_o(m_{lh}^{*}/m_o)e^{4}}{\pi\hbar^{3}\varepsilon^{2}}p_o\left(\frac{kT}{E_g}\right)^{7/2}\exp\left[-(1+\frac{m_{hl}^{*}}{m_{hh}^{*}})\frac{E_g}{kT}\right]g(\alpha) \tag{6}$$

In these equations, ε is the static dielectric constant, m_{hl}^{*} and m_{hh}^{*} are the effective masses of light and heavy holes, respectively, μ is the ratio of the electron to the heavy-hole effective mass. In the calculation of g_{A1} the product of the overlap integrals $|F_1F_2|$ is equal to 0.25 (Malyutenko, 1980). The calculated dependences are compared with experimental data published starting from the late 1950s to our days. As seen from Fig.7, the Auger 1 process is dominant at the electron concentration $n > 4\cdot10^{15}cm^{-3}$, whereas at $n < 1\cdot10^{15}cm^{-1}$ the carrier lifetime is determined by the radiative mechanism. In p-InSb the radiative recombination is dominant at the concentration $p < 5\cdot10^{15}$ cm^{-3}, Fig.8. The observed scatter of experimental data in samples with approximately the same concentration of carriers can be attributed to the SRH recombination. The values of the carrier lifetime in samples investigated earlier were lower than those obtained later. Thus, the relation between the improvement in technology of InSb and the increase in the carrier lifetime is clearly seen from experimental data shown in Fig. 7 and8.

In conclusion, the following peculiarities of the carrier lifetime in MWIR and LWIR HgCdTe may be pointed out: carrier lifetime can be essentially different in samples prepared by different growth techniques, even if they have approximately the same carrier concentration; correlation between the lifetime and the concentration of vacancies is observed in vacancy doped materials; lifetime can be increased by doping with foreign impurities; SRH recombination is more distinct at low temperatures and in samples with low carrier concentration.

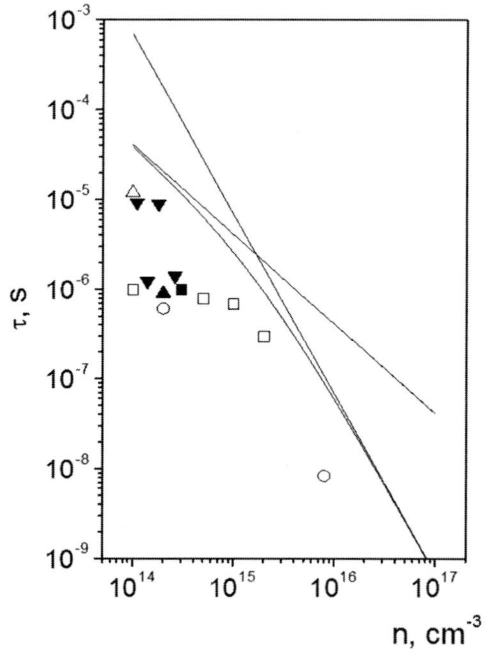

Figure 7. Calculated dependences (solid lines) of the lifetime in n-InSb at 77 K. Experimental are taken from (**Abduvakhidov, 1968**; **Malyutenko, 1980**; **Guseinov, 1971**; **Strelnikova, 1993**; **Biryulin, 2004**): open square, close square, open triangle, close triangle and open circle, respectively.

It seems that the SRH recombination is also dominant in InAs and InSb at low temperatures and low values of the carrier concentration. It should be noted that the low-temperature annealing of n-InSb leads to a significant increase in the lifetime in samples with the carrier concentration of the order of 10^{14} cm^{-3}. The highest values of the lifetime, as shown in Fig.6 (Strelnikova, 1993), were obtained in annealed samples due to significant decrease in the concentration of recombination centers.

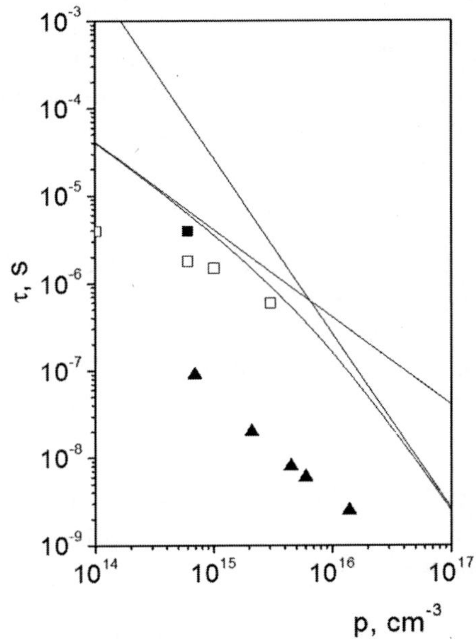

Figure 8. Calculated dependences (solid lines) of the lifetime in p-InSb at 77 K. Experimental data are taken from (Laff and Fan, 1961; **Volkov, 1967**; **Zitter, 1959**): close square, open square and close triangle, respectively.

5. CONCLUSION

Despite significant advances in the development of infrared photodiodes on narrow-gap II-VI and III-V semiconductors, theoretically predicted threshold parameters have not yet been achieved. The main reason for this is the participation of defects of different type in the processes of recombination and carrier transport. It is clear that further progress in development of IR photodiodes is closely connected with the band gap and defect engineering. The most impressive application of these concepts, based on knowledge of fundamental physical properties and defect states in narrow-gap semiconductors, is the development of HgCdTe planar heterostructure photodiodes with the highest performance achieved to-day.

REFERENCES

1. T. V. Abaeva, V. T. Bublik, A. N. Morozov, A. T. Pereverzev, 1987Effect of In and Sb vacancies on temperature dependence of InSb lattice parameter at high temperatures, Izv. Akad. Nauk SSSR, Neorg. Mater. (In Russia), 2320000-2337X.

2. J. M. Abduvakhidov, A. S. Volkov, V. V. Golovanov, 1968The study of the kinetics of photoconductivity and noise spectrum in InSb, Sov. Phys. Semicond., 210015-3222

3. E. Adomaytis, Z. Dorovolskis, A. Krotkus, 1984Picosecond photoconductivity of indium arsenide, Sov. Phys. Semicond., 1880015-3222

4. D. G. Adrianov, V. V. Karataev, G. V. Lazarev, et al.1977On the interaction of carriers with localized magnetic moments in InSb: Mn, Sov. Phys. Semicond.,1170015-3222

5. O. A. Ageev, A. E. Belyaev, N. S. Boltovets, V. N. Ivanov, R. V. Konakova, Ya. Kudryk, Ya, P. M. Lytvyn, V. V. Milenin, A. V. Sachenko, 2009Au-TiBx−n-6H-SiC Schottky barrier diodes: the features of current flow in rectifying and nonrectifying contacts, Semiconductors,4370015-3222

6. O. P. Agnihotri, H. C. Lee, K. Yang, 2002Plasma induced type conversion in mercury cadmium telluride, Semicond. Sci. Technol., 17R11-R19, 0268-1242

7. A. Ajisawa, N. Oda, 1995Improvement in HgCdTe Diode Characteristics by Low Temperature Post-Implantation Annealing, J. Electron. Mater., 2490361-5235

8. O. A. Allaberenov, N. V. Zotova, D. N. Nasledov, L. D. Neuimina, 1970Photoluminescence of n-InAs, Sov. Phys. Semicond.,4100015-3222

9. J. M. Arias, 1994Growth of HgCdTe by molecular beam epitaxy, in Properties of Narrow Gap Cadmium-Based Compounds, Capper P. (ed.), INSPEC, London, 30350-85296-880-9

10. J. M. Arias, J. G. Pasko, M. Zandian, L. J. Kozlowski, R. E. De Wames, 1994aMolecular beam epitaxy HgCdTe infrared photovoltaic detectors, Opt. Eng., 330091-3286

11. M. Arias, R. E. De Wames, S. H. Shin, J. G. Pasko, M. Chen, E. R. Gertner, 1989Infrared diodes fabricated with HgCdTe grown by molecular beam epitaxy on GaAs substrates, Appl. Phys. Lett. 5410251027110003-6951

12. M. Arias, J. G. Pasko, M. Zandian, S. H. Shin, G. M. Williams, L. O. Bubulac, R. E. De Wames, W. E. Tennant, 1993Planar p-on-n HgCdTe heterostructure photovoltaic detectors, Appl. Phys. Lett., 6290003-6951

13. V. P. Astahov, Yu. A. Danilov, V. F. Dutkin, V. P. Lesnikov, G. Sidorova, Yu, L. Suslov, A., I. I. Taubkin, Yu. M. Eskin, 1992Planar photodiodes based on InAs material, Techn. Phys. Lett. (In Russia), 1830320-0116

14. R. K. Bagai, G. L. Selth, W. N. Borle, 1983Growth of high purity indium antimony crystals for infrared detectors, Indian J. Pure Appl. Phys., 210019-5596

15. J. Bajaj, 2000State-of-the-art HgCdTe Infrared Devices, Proc. SPIE, 39480027-7786X.

16. I. M. Baker, C. D. Maxey, 2001Summary of HgCdTe 2D array technology in the U.K., J. Electron. Mater. 3060361-5235

17. L. A. Balagurov, E. M. Omel'yanovskii, V. I. Fistul', 1977The energy position of deep levels in high-resistance InAs: Cr, Sov. Phys. Semicond.,1120015-3222

18. A. Balderschi, N. O. Lipari, 1974Cubic contribution to the spherical model of shallow acceptor states, Phys. Rev. B., 941050-2947

19. A. N. Baranov, T. I. Voronina, A. A. Gorelenok, et al.1992Study of structural defects in epitaxial layers of indium arsenide, Semiconductors., 2690015-3222

20. A. N. Baranov, T. I. Voronina, T. S. Lagunova, et al.1993Properties of epitaxial indium arsenide doped with rare-earth elements, Semiconductors, 2730015-3222

21. N. L. Bazhenov, G. G. Zegrya, V. I. Ivanov-Omskii, 1997 Electroluminescence in the separated heterostructure of p-GaInAsSb/p- InAs at liquid helium temperatures, Semiconductors, 31100015-3222

22. A. R. Beattie, 1962Quantum Efficiency in InSb, J.Phys.Chem.Solids, 230022-3697

23. A. R. Beattie, P. T. Landsberg, 1959Auger effect in semiconductors, Proc. Roy. Soc. A., 2490308-2105

24. A. R. Beattie, G. Smith, 1967Recombination in semiconductors by a light hole Auger transition, Phys. Stat. Solidi, 190031-8965

25. K. W. Benz, G. Müller, 1979GaSb and InSb crystals grown by vertical and horizontal travelling heater method. J. Crystal. Growth, 460022-0248

26. M. Berding, M. van Schilfgaarde, A. Sher, 1994First-principles calculation of native defect densities in Hg0.8Cd0.2Te, Phys. Rev. B., 503151915340163-1829

27. M. A. Berding, A. Sher, M. van Schilfgaarde, 1995Defect modeling studies in HgCdTe and CdTe, J. Electron. Mat., 240361-5235

28. M. A. Berding, 2011Defects in HgCdTe-Fundamental, in Mercury cadmium telluride : growth, properties, and applications, Capper, P. and Garland, J. (eds.), Wiley, 263273978-0-47069-706-1

29. C. Besikci, 2000III-V infrared detectors on Si substrates, Proc. SPIE, 39480027-7786X

30. P. V. Biryulin, V. I. Turin, V. B. Yakimov, 2004Investigation of the characteristics of InSb photodiode arrays, Sov. Phys. Semicond., 3840015-3222

31. A. N. Blaut-Blachev, V. Ivlev, Selyanina. V. I. S.and, 1979Fluoride-fast diffusing acceptors in indium antimonide, Sov. Phys. Semicond., 13110015-3222

32. P. C. Boieriu, H. Grein, J. Garland, et al.2006Effects of hydrogen on majority carrier transport and minority carrier lifetimes in long-wavelength infrared HgCdTe on Si, J. Electron. Mater., 3560361-5235

33. R. Bornfreund, J. P. Rosbeck, Y. N. Thai, E. P. Smith, D. D. Lofgreen, M. F. Vilela, A. A. Buell, M. D. Newton, K. Kosai, S. M. Johnson, T. J. De Lyon, J. J. Jensen, M. Z. Tidrow, 2007High-Performance LWIR MBE-Grown HgCdTe/Si Focal Plane Arrays, J. Electron. Mater., 370361-5235

34. V. T. Bublik, A. P. Blaut-Blachev, V. V. Karataev, M. G. Mil'vidskii, et al.1977Nature of intrinsic point defects in indium arsenide and their effect on electrophysical proiperties of single crystals, Kristalografiya (Sov. Phys. Crystallogr.), 2260023-4761

35. V. T. Bublik, V. V. Karataev, M. G. Mil'vidskii, et al.1979Defects in heavily doped with donor impurities of Group VI single crystals of indium arsenide, Kristalografiya (Sov. Phys. Crystallogr.), 2430023-4761

36. V. T. Bublik, V. V. Karataev, M. G. Mil'vidskii, et al.1979aDefects in heavily doped with tin single crystals of InAs, Kristalografiya (Sov. Phys. Crystallogr.), 2450023-4761

37. L. O. Bubulac, 1988Defects, diffusion and activation in ion implanted HgCdTe, J. Cryst. Growth., 861-40022-0248

38. M. A. Bynin, Yu. A. Matveev, 1985Electronic structure of anion vacancies in indium arsenide, Sov. Phys. Semicond., 19110015-3222

39. L. C. Cai, H. Chen, L. , Q. Huang, J. M. Zhou, 2003Raman spectroscopic studies of InAs epilayers grown on the GaAs (001) substrates, J. Crystal Growth, 253ISSN 0022-0248

40. P. Capper, C. T. Elliot, Eds, 2001Infrared Detectors and Emitters: Material and Devices, Kluwer Academic Publishers, 0-79237-206-9

41. P. Capper, 1991A review of impurity behavior in bulk and epitaxial Hg1-xCdxTe, J. Vac. Sci. Technol. B, 930073-4211X.

42. P. Capper, J. Garland, Eds, 2011Mercury Cadmium Telluride: Growth, Properties, and Applications, Wiley 2011, 978-0-47069-706-1

43. P. Capper, Ed, 1994Properties of Narrow Gap Cadmium-based Compounds, INSPEC, London, 0-85296-880-9

44. P. Capper, Ed, 1997Narrow-gap II-VI Compounds for Optoelectronic and Electromagnetic Applications, Chapman and Hall, London, 1997, 0-41271-560-0

45. T. N. Casselman, 1981Calculation of the Auger lifetime in p-type Hg1-xCdxTe, J. Appl. Phys., 520021-8979

46. T. N. Casselman, P. E. Petersen, 1980A comparison of the dominant Auger transitions in p-type (Hg,Cd)Te, Solid State Commun., 330038-1098

47. H. Chen, L. C. Cai, C. L. Bao, J. H. Li, Q. Huang, J. M. Zhou, 2000Two-step method to grow InAs epilayer on GaAs substrate using a new prelayer, J. Crystal Growth, 2081-40022-0248

48. D. T. Cheung, 1985An overview on defect studies in MCT, J. Vac. Sci. Technol., A310734-2101

49. J. Chu, A. Sher, 2008Physics And Properties of Narrow Gap Semiconductors, Springer, 978-0-38774-743-9

50. J. Chu, A. Sher, 2010Device Physics of Narrow Gap Semiconductors, Springer, 978-1-44191-039-4

51. R. Fastow, D. Goren, Y. Nemirovsky, 1990Shockley-Read recombination and trapping in p-type HgCdTe, J. Appl. Phys., 6870021-4651

52. A. Dixit, B. Bansal, V. Venkataraman, G. N. Subbanna, K. S. Chandrasekharan, B. M. Arora, H. L. Bhat, 2002High-mobility InSb epitaxial films grown on a GaAs(001) substarte using liquid-phase epitaxy, Appl. Phys. Lett., 800003-6951

53. V. A. Dixit, B. V. Rodrigues, R. Venkataraman, K. S. Chandrasekharan, K. S. Chandrasekharan, B. M. Arora, H. L. Bhat, 2002aGrowth of InSb epitaxial layer on GaAs(001) substrate by LPE and their characterizations, J. Cryst. Growth, 2350022-0248

54. W. Dobbelaere, J. Boech, R. Heremans, et al.1992InAs p- n diodes grown on GaAs and GaAs-coated Si by molecular beam epitaxy, Appl. Phys. Lett, 6070003-6951

55. G. Eftekhari, 1997The Effect of Sulfur Passivation and Rapid Thermal Annealing on the Properties of InAs MOS Structures with the Oxide Layer Deposited by Reactive Sputtering, Phys. Stat. Solidi (a), 16120031-8965

56. R. J. Egan, T. L. Tansley, V. W. L. Chin, 1995Growth of InAs from monoethyl arsine, J. Crystal Growth, 1471-20022-0248

57. S. Egemberdieva, Sh, S. D. Luchinin, T. Saysenbaev, et al.1982Deep levels in the band gap of indium antimonide, Sov. Phys. Semicond., 1630015-3222

58. N. P. Esina, N. V. Zotova, B. A. Matveev, et al.1985Features of the luminescence of plastically deformed heterostructures of InAsSbP/ InAs, Sov. Phys. Semicond.,19110015-3222

59. V. V. Evstropov, M. Dzhumaeva, Yu. V. Zhilyaev, N. Nazarov, A. A. Sitnikova, L. M. Fedorov, 2000Dislocation origin and a model of the excessive tunnel current in GaP p-n structures, Semiconductors, 34110015-3222

60. V. V. Evstropov, Yu. V. Zhilyaev, M. Dzhumaeva, N. Nazarov, 1997Tunnel excess current in nondegenerated (p-n and m-s) silicon-containing III-V compound semiconductor structures, Semiconductors, 3120015-3222

61. S. Farrell, Mulpuri. Rao, G. Brill, Y. Chen, P. Wijewarnasuriya, N. Dhar, D. Benson, K. Harris, 2011Effect of Cycle Annealing Parameters on Dislocation Density Reduction for HgCdTe on Si, J. Electron. Mater., 4080361-5235

62. I. A. Fomin, L. V. Lebedeva, N. M. Annenko, 1984Investigation of deep levels in InAs using capacitance measurements of MIS structures, Sov. Phys. Semicond.,1840015-3222

63. A. M. Fowler, I. Gatley, P. Mc Intyre, F. J. Vrba, A. Hoffman, 1996ALADDIN, the 1024-1024 InSb array: design, description, and results, Proc.SPIE, 28160027-7786X

64. H. Fujisada, M. Kawada, 1985Temperature Dependence of Reverse Current in Be Ion Implanted InSb p+n Junctions, J. Appl. Phys., 24L76-L78, 0021-8979

65. T. Fukui, Y. Horikoshi, 1979Organometallic VPE Growth of InAs, Jpn. J. Appl. Phys., 181347-4065

66. T. I. Galkina, N. A. Penin, V. A. Rassushin, 1966Determination of the energy of the acceptor level of cadmium in indium arsenide, Sov.Phys. Solid State, 880367-3294

67. H. H. Gao, A. Krier, V. V. Scherstnev, 1999High quality InAs growth by liquid phase epitaxy using gadolinium gettering, Semicond. Sci. Technol., 1430268-1242

68. J. M. B. E. Garland, of. Growth, Cadmium. Mercury, pp.131-14. Telluride, Mercury. in, telluride. cadmium, properties. growth, Capper. applications, and, J. MBE Growth of Mercury Cadmium Telluride, 131149in Mercury cadmium telluride: growth, properties, and applications, Capper, P. and Garland, J. (Eds.), Wiley, 978-0-47069-706-1

69. J. Garland, R. Sporken, 2011Substrates for the Epitaxial Growth of MCT, 7594in Mercury cadmium telluride: growth, properties, and applications, Capper, P. and Garland, J. (Eds.), Wiley, 978-0-47069-706-1

70. B. L. Gelmont, 1978Three-Band Kane Model of Auger Recombination, JETP (In Russia), 75N2, 536-544, 0044-4510

71. B. L. Gelmont, 1981Auger Recombination in Narrow-Gap p-Type Semiconductor, Sov. Phys. Semicond., 15N7, 1316-1319, 0015-3222

72. B. L. Gelmont, Z. N. Sokolova, 1982Auger Recombination in Direct-Gap n-Type Semiconductors, Sov. Phys. Semicond., 16N9, 1670-1672, 0015-3222

73. B. L. Gelmont, Z. N. Sokolova, I. N. Yassievich, 1982Auger Recombination in Direct-Gap p-Type Semiconductors, Sov. Phys. Semicond., 16N3, 592-600, 0015-3222

74. E. I. Gheorghitse, I. T. Postolani, V. A. Smirnov, P. G. Untila, 1989Photoluminescence of p-InAs:Mn, Sov. Phys. Semicond., 234 0015-3222

75. T. D. Golding, O. W. Holland, M. J. Kim, J. H. Dinan, L. A. Almeida, J. M. Arias, J. Bajaj, H. D. Shih, W. P. Kirk, 2003HgCdTe on Si: present status and novel buffer layer concepts, J. Electron. Mater., 3280361-5235

76. V. V. Golovanov, V. G. Oding, 1969The influence of deep-level compensation on the electrical properties of p-InSb, Sov. Phys. Semicond., 320015-3222

77. V. V. Golovanov, E. L. Ivchenko, V. G. Oding, 1973Generation-recombination noise in p-InSb at 78 K, Sov. Phys. Semicond., 740015-3222

78. A. Granato, K. Lücke, 1956Theory of Mechanical Damping Due to Dislocations, J. Appl. Phys.,2760021-8979

79. E. K. Guseinov, R. I. Ibragimov, V. G. Korotin, D. N. Nasledov, Yu. G. Popov, 1971The recombination processes in n-InSb in the temperature range 4.2- 77 K, Sov. Phys. Semicond., 590015-3222

80. E. K. Guseinov, M. P. Mikhailova, D. N. Nasledov, et al.1969Impurity photoconductivity in InAs, Sov. Phys. Semicond.,3110015-3222

81. M. I. Guseva, N. V. Zotova, A. V. Koval', D. N. Nasledov, 1975Radiative recombination in indium arsenide implanted with Group IV elements, Sov. Phys. Semicond.,950015-3222

82. Halt, D.B. and Yacobi, G.Ya2007Structural Defects in Semiconductors. Electronic Properties, Device Effects and Structures, Cambridge University Press, 0-52181-934-2

83. S. K. Haywood, R. W. Martin, N. J. Mason, P. J. Walker, 1990Growth of InAs by MOVPE using TBAs and TMIn, J. Electron. Mater., 1980361-5235

84. W. He, Z. Celik-Batler, 1996f noise and dark current components in HgCdTe MIS infrared detectors, Solid-State Electron., 1910038-1101

85. A. W. Hoffman, D. Randall, 1991High-performance 256 x 256 InSb FPA for astronomy, Proc. SPIE, 15400027-7786X.

86. J. E. L. Hollis, S. C. Choo, E. L. Heasell, 1967Recombination center in InSb, J. Appl. Phys., 3840021-8979

87. D. E. Holmes, G. S. Kamath, 1980Growth-characteristics of LPE InSb and InGaSb, J. Electron. Mater., 90361-5235

88. D. B. Holt, B. G. Yacobi, 2007Extended defects in Semiconductors. Electronic Properties, Device Effects and Structures, Cambridge University Press, 978-0-52181-934-3

89. K. T. Huang, Y. Hsu, R. M. Cohen, G. B. Stringfellow, 1995OMVPE growth of InAsSb using novel precursors, J.Crystal Growth, 15640022-0248

90. K. F. Hulme, J. B. Mullin, 1962Indium Antimonide-A review of its Preparation, Properties and Device Applications, Solid-State Electron., 50038-1101

91. C. E. Hurwitz, J. P. Donnelly, 1975Planar InSb Photodiodes Fabricated by Be and Mg Ion Implantation, Solid State Electron., 180038-1101

92. M. I. Iglitsyn, E. Solovyov, 1968Determination of the ionization energy of cadmium in indium arsenide, Sov. Phys. Semicond., 270015-3222

93. J. A. Ilyenkov, T. E. Kovalevskaya, A. P. Kovchavtsev, 1992Estimation of the parameters of deep levels in MIS structures based on InAs, Poverhnost: fizika, himiya, mehanika (In Russia), 1162690734-1520

94. N. M. Ismailov, D. N. Nasledov, Y. S. Smetannikova, 1969The impurity photoconductivity of indium antimonide at low temperatures, Sov. Phys. Semicond.,26

95. Z. F. Ivasiv, F. Sizov, Tetyorkin. V. V. F.and, 1999Noise spectra and dark current investigations n+-p type Hg1-xCdxTe (x0.22) photodiodes, Semicond. Phys. Quant. Electron. Optoelectron. (Kiev), 231605-6582

96. S. M. Johnson, D. R. Rhiger, J. Rosbeck, P. Peterson, J. M. , S. M. Taylor, M. E. Boyd, 1992Effect of dislocations on the electrical and optical properties of long-wavelength infrared HgCdTe photovoltaic detectors, J. Vac. Sci. Technol. B, 100734-2101

97. C. E. Jones, V. Nair, J. Lindquist, D. L. Polla, 1982Effects of deep-level defects in Hg1−xCdxTe provided by DLTS, J. Vac. Sci. Technol. 2110734-2101

98. A. Jóźwikowska, K. Jóźwikowski, J. Rutkowski, Z. Orman, A. Rogalski, 2004Generation-recombination effects in high temperature HgCdTe heterostructure photodiodes, Opto-Electron. Rev., 1241230-3402

99. S. Kalem, J. Chyi-I, H. Morkoç, R. Bean, K. Zanio, 1998Growth and transport properties of InAs epilayers on GaAs, Appl. Phys. Lett., 53170003-6951

100. V. V. Karataev, G. A. Nemtsova, N. S. Rizhova, T. G. Yugova, 1977Effect of heat treatment on the electrical properties of undoped indium arsenide, Sov. Phys. Semicond., 1190015-3222

101. K. Kazaki, A. Yahata, W. Miyao, 1976Properties of InSb Photodiodes Fabricated by Liquid Phase Epitaxy, J. Appl. Phys., 150021-8979

102. F. P. Kesamanly, T. S. Lagunova, D. N. Nasledov, et al.1968Electrical properties of p-type indium arsenide crystals, Sov. Phys. Semicond., 210015-3222

103. M. N. Kevorkov, A. N. Popkov, V. S. Uspensky, et al.1980Thermal acceptors in indium antimonide, Izv. Akad. Nauk SSSR, Neorg. Mater. (In Russia), 16120000-2337X.

104. I. Kimukin, N. Biyikli, E. Ozbay, 2003InSb high-speed photodetectors grown on GaAs substrate, J. Appl. Phys., 94, 151554165414O021-8979

105. Kinch M.A.1981Metal-insulator semiconductor infrared detectors, in Semiconductor and Semimetals, 18Willardson, R.K. and Beer, A.C. (Eds.), New York: Academic Press, ch.7, 978-0-12752-118-3

106. Kornyushkin NA, NA Valisheva, Kovchavtsev AP, GL Kuryshev1996Influence of the interface and deep levels in the forbidden gap on the capacitance-voltage characteristics of InAs MIS structures, Semiconductors.,3050015-3222

107. V. G. Korotin, S. N. Krivonogov, D. N. Nasledov, Y. S. Smetannikova, 1976The model of recombination processes in n-InSb, Sov. Phys. Semicond., 1010015-3222

108. O. V. Kosogov, L. S. Perevyaskin, 1970Electrical Properties of Epitaxial p+-n Junctions in Indium Antimonide, Sov. Phys. Semicond., 80015-3222

109. L. Kozlowski, K. Vural, J. Luo, A. Tomasini, T. Liu, W. K. Kleinhans, 1999Low-noise infrared and visible focal plane arrays, Opto-Electron. Rev., 71230-3402

110. S. Krishnamurthy, M. A. Berding, H.and. Robinson, A. Sher, 2006 Tunneling in long-wavelength infrared HgCdTe photodiodes, J. Electron. Mater., 3560361-5235

111. C. H. Kuan, R. M. Lin, S. F. Tang, T. P. Sun, 1996Analysis of the Dark Current in the Bulk of InAs Diode Detectors, J. Appl. Phys., 800021-8979

112. M. Kumagawa, A. F. Witt, M. Lichtenstelger, H. C. Gatos, 1973Current-controlled and dopant modulation in liquid phase epitaxy, J. Electrochem. Soc., 1200013-4651

113. G. L. Kuryshev, A. P. Kovchavtsev, N. Valisheva, 2001Electronic properties of MIS structures based on InAs, Semiconductors, 3590015-3222

114. G. L. Kuryshev, I. I. Lee, V. M. Bazovkin, et al.2009Threshold parameters of multielement InAs hybrid IR FPA and InAs-based devices, Prikladnaya Fizika (In Russia), 2279921996-0948

115. N. Kuze, E. G. Camargo, K. Ueno, et al.2007High performance miniaturized InSb photovoltaic infrared sensors operating at room temperature, J. Crystal Growth, 301-3020022-0248

116. V. Kveder, M. Kittler, W. Schröter, 2001Recombination activity of contaminated dislocations in silicon: A model describing electron-beam-induced current contrast behavior, Phys. Rev.B, 630163-1829

117. V. V. Kveder, R. Labusch, Yu. A. Ossipyan, 1985Frequency dependence of the dislocation conduction in Ge and Si, Phys. Stat. Sol., 920370-1972

118. R. Labusch, 1982One dimensional transport along dislocations, Physica, 117B-118B10921-4526

119. R. Labusch, W. Schröter, 1983Electrical Properties of Dislocations in Semiconductors, in Dislocations in Solids, Nabarro, F.R.N. (ed.), 5Amsterdam: North-Holland, 0-44485-050-3

120. Laff R.A. and Fan H.Y.1961Carrier lifetime in indium antimonide, Phys. Rev., 1211

121. S. Liang, 1966Preparation of Indium Antimonide, in Compound Semiconductors, Willardson, R.K. and Georing, H.L. (Eds.), 227237N.Y.

122. R. M. Lin, S. F. Tang, S. C. Lee, C. H. Kuan, G. S. Chen, T. P. Sun, Wu. J. C. And, 1997Room Temperature Unpassivated InAs p-i-n

Photodetectors Grown by Molecular Beam Epitaxy, IEEE Trans. Electron Dev., 440018-9383

123. C. L. Litter, D. G. Seiler, M. R. Loloee, 1990Magneto-optical investigations of impurity and defect levels in HgCdTe alloys, J. Vac. Sci. Technol.A, 820734-2101

124. W. K. Liu, J. Winesett, Xuemei. Weiluan, Santos. M. B. Zhang, X. M. Fang, P. J. Mc Cann, 1997Molecular beam epitaxy of InSb on Si substrates using fluoride buffer layers, J. Appl. Phys., 8140021-8979

125. V. C. Lopes, A. J. Syllaios, M. C. Chen, 1993Minority carrier lifetime in mercury cadmium telluride, Semicond. Sci. Technol., 820268-1242

126. P. Madejczyk, A. Piotrowski, W. Gawron, K. Klos, J. Pawluczyk, J. Rutkowski, J. Piotrowski, J. , A. Rogalski, 2005Growth and properties of MOCVD HgCdTe epilayers on GaAs substrates, Opto-Electron. Rev., 1331230-3402

127. O. Madelung, 1996Semiconductors-Basic Data, 2nd revised Edition, Springer, 3-54060-883-4

128. O. Madelung, U. Rössler, M. . Schulz, Eds, 2003Landolt-Börnstein- Group III Condensed Matter. Numerical Data and Functional Relationships in Science and Technology, 41A2b Impurities and Defects in Group IV Elements, IV-IV and III-V Compounds. Part b: Group IV-IV and III-V Compounds, Springer, 978-3-54043-086-5

129. J. Mahony, P. Maseher, 1977Position- annihilation study of vacancy defects in InAs, Phys. Rev. B., 1997, 55150015-3222

130. H. Maier, J. Hesse, 1980Growth, properties and applications of narrow-gap semiconductors, in Crystall Growth Properties and Applications, Freyhard, H.C. (Ed.), Springer Verlag, Berlin, 145219

131. V. N. Malyutenko, S. S. Bolgov, V. I. Pipa, V. I. Chaykin, 1980Quantum yield of recombination radiation in n-InSb, Sov. Phys. Semicond., 14N4, 781-786, 0015-3222

132. K. D. Maranowski, J. M. Peterson, S. M. Johnson, J. B. Varesi, W. A. Radford, A. C. Chields, R. E. Bornfreund, A. A. Buell, 2001MBE

growth of HgCdTe on silicon substrates for large format MWIR focal plane arrays, J. Electron. Mater., 300361-5235

133. H. F. Matare, 1971Defect Electronics in Semiconductors, Wiley, ISBN 139780471576181

134. C. D. Maxey, 2011Metal-Organic Vapor Phase Epitaxy (MOVPE) Growth, in Mercury cadmium telluride: growth, properties, and applications, Capper, P. and Garland, J. (Eds.), 113129Wiley, 978-0-47069-706-1

135. P. J. Mc Nall, 1970Ion Implantation in InAs and InSb, in Radiation Effects and Defects in Solids, 616747348

136. I. Mengailis, A. R. Calawa, 1966Solution regrowth of planar InSb lase structure, J. Electrochem. Soc., 1130013-4651

137. W. F. M. Micklethwaite, A. J. Johnson, 2000InSb: materials and devices, in Infrared Detectors and Emitters: Materials and Devices, Capper, P. and Elliott, C.T. (Eds.), Kluwer Academic Publishers, Boston, 177204978-0-79237-206-6

138. Milnes A.G.1973Deep impurities in semiconductors, Wiley, 0-47160-670-7

139. R. L. Mozzi, J. M. Lavine, 1970Zn-Diffusion Damage in InSb Diodes, J. Appl. Phys., 410021-8979

140. J. A. Mroczkowski, J. F. Shanley, M. B. Reine, Vecchio. P. Lo, D. L. Polla, 1981Lifetime measurement in Hg0.7Cd0.3Te by population modulation, Appl. Phys. Lett., 3840003-6951

141. D. N. Nasledov, Y. S. Smetannikova, 1962Temperature dependence of the lifetime of carriers in indium antimonide, Sov. Phys. Solid State, 410367-3294

142. Y. Nemirovsky, A. Unikovsky, 1992Tunnelling and 1/f noise currents in HgCdTe photodiodes, J. Vac. Sci. Technol. B., 1040734-2101

143. Y. Nemirovsky, R. Fastow, M. Meyassed, A. Unikovsky, 1991Trapping effects in HgCdTe, J.Vac.Sci.Technol. B., 930734-2101

144. Y. Nemirovsky, D. Rosenfeld, R. Adar, A. Kornfeld, 1989Tunneling and dark currents in HgCdTe photodiodes, J.Vac.Sci.Technol. A, 720734-2101

145. K. Nishitani, K. Nagahama, T. Murotani, 1983Extremely Reproducible Zinc Diffusion into InSb and Its Application to Infrared Detector Array, J. Electron. Mater., 1210361-5235

146. R. Nitccki, B. Pohoryles, (1985, 1985Tunneling from dislocation cores in silicon Schottky diodes, Appl. Phys., A360947-8396

147. P. Norton, 2002HgCdTe infrared detectors, Opto-Electron. Rev., 1031230-3402

148. P. R. Norton, 1999Infrared detectors in the next millennium, Proc.SPIE, 36980027-7786X.

149. E. M. Omel'yanovskii, V. I. Fistul', L. A. Balagurov, et al.1975On the behavior of transition-metal impurities in III-V compounds, Sov. Phys. Semicond., 930015-3222

150. Yu. A. Osipiyan, I. B. Savchenko, 1968Experimental observation of the influence of light on plastic deformation of cadmium sulphide, JETP Letters (In Russia), 70021-3640

151. S. Ozer, C. Besikci, 2003Assessment of InSb photodetectors on Si substrates, J. Phys. D: Appl. Phys., 50022-3727

152. S. G. Parker, O. W. Willson, B. H. Barbel, 1965Indium antimonide of high perfection, J. Electrochem. Soc., 1120013-4651

153. W. J. Parrish, J. D. Blackwell, G. T. Kincaid, R. C. Paulson, 1991Low-cost high-performance InSb 256 x 256 infrared camera, Proc. SPIE, 15400027-7786X.

154. D. L. Partin, L. Green, D. T. Morelli, J. Heremans, B. K. Fuller, C. M. Thrush, 1991Growth and characterization of indium arsenide thin films, J. Electron. Mater., 20120361-5235

155. J. Pehek, H. Levinstein, 1965Recombination radiation from InSb. Phys. Rev., 1402576586

156. P. E. Petersen, 1970Auger Recombination in Hg1-xCdxTe, J. Appl. Phys., 410021-4922

157. P. E. Petersen, 1981Auger Recombination in Mercury Cadmium Telluride, in Semiconductors and Semimetals, 18Willardson, R.K. and Beer A.C. (Eds.), Academic Press, 978-0-12752-118-3

158. J. M. Peterson, J. A. Franklin, M. Readdy, S. M. Johnson, E. Smith, W. A. Radford, I. Kasai, 2006High-quality large-area MBE HgCdTe/Si, J. Electron. Mater., 360361-5235

159. A. Plitnikas, A. Krotkus, Z. Dorovolskis, 1982The current-voltage characteristics of compensated indium arsenide in strong electric fields, Sov. Phys. Semicond.,1660015-3222

160. D. L. Polla, C. E. Jones, 1981Deep level studies of Hg1−xCdxTe. I: Narrow-band-gap space-charge spectroscopy, J. Appl. Phys., 5280021-8979

161. D. L. Polla, R. L. Aggarwal, J. A. Mroczkowski, J. F. Shanley, M. B. Reine, 1982Observation of deep levels in Hg1−xCdxTe with optical modulation spectroscopy, Appl. Phys. Lett., 4040003-6951

162. D. L. Polla, M. B. Reine, C. E. Jones, 1981aDeep level studies of Hg1−xCdxTe. II: Correlation with photodiode performance, J. Appl. Phys., 5280021-8979

163. D. L. Polla, S. P. Tobin, M. Reine, Sood. A. K. B.,and, 1981bExperimental determination of minority-carrier lifetime and recombination mechanisms in p-type Hg1−xCdxTe, J. Appl. Phys., 5280021-8979

164. M. Razeghi, 2003Overview of antimonide based III-V semiconductor epitaxial layers and their applications at the center for quantum devices, Eur. Phys. J. Appl. Phys. 231286-0042

165. M. B. Reine, 2000Photovoltaic detectors in MCT, In Infrared Detectors and Emitters: Materials and Devices, P.Capper and C.T. Elliott, Eds., Kluwer Academic Publishers, Boston, 0-79237-206-9

166. A. Rogalski, 2009Infrared detectors for the future, Acta Physica Polonica A, 11630587-4246

167. A. Rogalski, 2011Infrared Detectors, 2nd ed., CRC Press, 978-1-42007-671-4

168. A. Rogalski, K. Adamiec, J. Rutkowski, 2000Narrow-Gap Semiconductor Photodiodes, SPIE Press, Bellingham, 0-81943-619-4

169. A. Rogalski, Ed, 1995Infrared Photon Detectors, SPIE Optical Engineering Press, 081941798

170. J. P. Rosbeck, I. Kassi, R. M. Hoendervoog, T. Lanir, 1981High Performance Be Implanted InSb Photodiodes, IEEE IEDM, 811074-1879

171. D. Rosenfeld, G. Bahir, 1992A model for the trap-assisted tunnelling mechanism in diffused n-p and implanted n+-p HgCdTe photodiodes, IEEE Trans. Electron. Dev., 3970018-9383

172. H. R. Schaake, 1985The existence region of the Hg0.8Cd0.2Te phase field, J. Electron. Mater., 1450361-5235

173. H. R. Schaake, J. H. Tregilgas, A. J. Lewis, M. Everett, 1983Lattice defects in (Hg,Cd)Te: Investigations of their nature and evolution, J. Vac. Sci. Technol. A, 130734-2101

174. M. Seibt, R. Halil, V.and. Kveder, W. Schröter, 2009Electronc states at dislocations and metal silicide precipitates in crystalline silicon and their role in solar cell materials, J. Appl. Phys A, 960021-8979

175. D. Shaw, P. Capper, 2011Extrinsic Doping, in Mercury cadmium telluride: growth, properties, and applications, Capper, P. and Garland, J. (Eds.), 317337Wiley, 978-0-47069-706-1

176. O. S. Shepelina, Y. F. Novototsky-Vlasov, 1992Equilibrium parameters of deep levels in bulk indium antimonide, Semiconductors, 2660015-3222

177. A. Sher, M. A. Berding, M. van Schilfgaarde, An. Chen-Ban, 1991HgCdTe status review with emphasis on correlations, native defects and diffusion, Semicond. Sci. Technol., 6C59-C70, 0268-1242

178. V. B. Shikin, Yu. V. Shikina, 1995Charged dislocations in semiconductors, Physics-Uspekhi (Advances in Physical Sciences), In Russia, 3880042-1294

179. S. H. Shin, J. M. Arias, D. D. Edwall, M. Zandian, J. G. Pasko, R. E. De Wammes, 1992Dislocation reduction in HgCdTe in GaAs and Si, J. Vac. Sci. Technol. B, 100022-5355

180. S. H. Shin, J. M. Arias, M. Zandian, J. G. Pasko, R. E. De Wames, 1991Effect of the dislocation density on minority-carrier lifetime in molecular beam epitaxial HgCdTe, Appl. Phys. Lett., 590003-6951

181. M. A. Sipovskaya, Y. S. Smetannikova, (1984, 1984The dependence of the carrier lifetime on the electron density in n-InSb, Sov. Phys. Semicond., 1820015-3222

182. I. A. Strelnikova, N. G. Ermakov, A. V. Laptev, M. R. Rauhman, 1993Effect of heat treatment on the properties of indium antimonide, Inorg. Mater. (In Russia), 2930000-2337X.

183. A. Sukach, V. Tetyorkin, G. Olijnyk, V. Lukyanenko, A. Voroschenko, 2005Cooled InAs photodiodes for IR applications, Proc. SPIE., 59570027-7786X.

184. Sukach A.V. and Tetyorkin V.V.2009Ultrasonic treatment-induced modification of the electrical properties of InAs p-n junctions, Tech. Phys. Lett., 36N6, 514-517, 1063-7850

185. A. Sukach, V. Tetyorkin, G. Olijnuk, V. Lukyanenko, A. Voroschenko, 2005Cooled InAs photodiodes for IR applications, Proc. SPIE, 59570027-7786X.

186. S. M. Sze, 1981Physics of Semiconductor Devices, second edition, Wiley, N.Y. 0-47105-661-8

187. V. Tetyorkin, A. Sukach, A. Tkachuk, 2011InAs Infrared Photodiodes, in Advances in Photodiodes, Dalla Betta, G-F. (ed.), Intech Open Acces Publisher, 978-9-53307-163-3

188. J. H. Tregilgas, T. L. Polgreen, M. C. Chen, 1988Dislocations and electrical characteristics of HgCdTe, J. Crystal Growth, 861-40022-0248

189. P. Tribolet, P. Chorier, F. Pistone, 2003Key performance drivers for coded large IR staring arrays, Proc. SPIE, 50740027-7786X.

190. V. I. Trifonov, N. G. Yaremenko, 1971The deep donor level in n-InSb, Sov. Phys. Semicond., 550015-3222

191. N. P. Tsitsina, A. P. Fadeeva, E. E. Vdovkina, et al.1975Effect of low-temperature annealing on the properties of InSb, Izv. Akad. Nauk SSSR, Neorg. Mater (In Russia), 1150000-2337X.

192. E. G. Valyashko, T. B. Pleskacheva, N. D. Tyapkina, 1975Effect of heat treatment on electrical properties and impurity photoconductivity of p-InSb, Izv. Akad. Nauk SSSR, Neorg. Mater., (In Russia), 1160000-2337X.

193. A. S. Volkov, V. V. Golovanov, 1967Recombination processes in p-InSb, Sov. Phys. Semicond., 120015-3222

194. Eichel. von-Streiber, C. , M. Behet, M. Heuken, K. Heime, 1997Doping of InAs, GaSb and InPSb by low pressure MOVPE, J. Crystal Growth, 1701-40022-0248

195. T. I. Voronina, T. S. Lagunova, S. S. Kizhaev, et al.2004Growth and magnesium doping of InAs layers by vapor-phase epitaxy from organometallic compounds, Semiconductors, 3850015-3222

196. T. I. Voronina, T. S. Lagunova, K. D. Moiseev, et al.1999Electrical properties of epitaxial indium arsenide and narrow-gap solid solutions on its base, Semiconductors, 3370015-3222

197. T. I. Voronina, N. V. Zotova, S. S. Kizhaev, 1999aFluorescent and other properties of InAs layers and p-n-structures on their base, grown by vapor-phase epitaxy from organometallic compounds, Semiconductors, 33100015-3222

198. H. R. Vydyanath, 1981Lattice Defects in Semiconducting Hg1−xCdxTe Alloys, J. Electrochem. Soc., 128120013-4651

199. H. R. Vydyanath, 1991Mechanisms of incorporation of donor and acceptor dopants in (Hg,Cd)Te alloys, J. Vac. Sci. Technol. B, 90734-2101

200. J. Y. Wang, 1980Effect of trap tunneling on the performance of long-wavelength Hg1-xCdxTe photodiodes, IEEE Trans. on Electron Dev., ED-2710018-9383

201. S. P. Watkins, C. A. Tran, R. Ares, G. Soerensen, 1995High mobility InAs grown on GaAs substrates using tertiarybutyl arsine and trimethylindium, Appl. Phys. Lett., 6670003-6957

202. M. Whelan, 1969Leakage currents of n/p silicon diodes with different amounts of dislocations, Solid-State Electronics, 1260038-1101

203. J. T. Wimmers, D. S. Smith, 1983Characteristics of InSb photovoltaic detectors at 77 K and below, Proc.SPIE, 3640027-7786X.

204. J. T. Wimmers, R. M. Davis, C. A. Niblack, D. S. Smith, 1988Indium antimonide detector technology at Cincinnati Electronics Corporation, Proc.SPIE, 9300027-7786X.

205. T. Yamamoto, Y. Miyamoto, K. Tanikawa, 1985Minority carrier lifetime in the region close to the interface between the anodic oxide CdHgTe, J. Crystal Growth, 7210022-0248

206. J. Yang, Z. Yu, D. Tang, 1985The defects in Hg0.8Cd0.2Te annealed at high temperature, J. Crystal Growth, 721-20022-0248

207. M. Yano, M. Nogami, Y. Matsuchima, M. Kimata, 1977Molecular beam epitaxial growth of InAs, Japan. J. Appl. Phys., 16120021-4922

208. I. Yonenaga, 1998Dynamic behavior of dislocations in InAs: in comparison with III-V compounds and other semiconductors, J. Appl. Phys., 840021-4922

209. F. A. Zaitov, O. V. Gorshkov, A. Y. Polyakov, et al.1981The nature of deep acceptors in indium antimonide, Sov. Phys. Semicond., 1560015-3222

210. R. N. Zitter, A. J. Strauss, A. E. Attard, 1959Recombination processes in p-type indium antimonide, Phys. Rev., 1152

211. N. V. Zotova, V. V. Karataev, A. V. Koval', 1975 Photoluminescence of n-InAs crystals doped with tin, Sov. Phys. Semicond.,9100015-3222

10

CHARACTERIZATION OF A PHOTODIODE COUPLED WITH A SI NANOWIRE-FET ON A PLASTIC SUBSTRATE

Kiyeol Kwak [1], Kyoungah Cho [1] and Sangsig Kim [1,2,*]

[1] Department of Electrical Engineering, Korea University, Seoul 136-713, Korea

[2] Department of Nano Semiconductor Engineering, Korea University, Seoul 136-713, Korea

ABSTRACT

In this study, a laterally coupled device composed of a photodiode and a Si nanowires-field-effect transistor (NWs-FET) is constructed on a plastic substrate and the coupled device is characterized. The photodiode is made of *p*-type Si NWs and an *n*-type ZnO film. The Si NWs-FET is connected electrically to the photodiode in order to enhance the latter's photocurrent efficiency by adjusting the gate voltage of the FET. When the FET is switched on by biasing a gate voltage of −9 V, the photocurrent efficiency of the photodiode is three times higher than that when the FET is switched off by biasing a gate voltage of 0 V.

KEYWORDS

pn heterojunction photodiode; FET; optical sensor; plastic substrate; nanowire

1. INTRODUCTION

Inorganic nanomaterial-based plastic optoelectronics is a rapidly developing field, since inorganic nanomaterials can be easily fabricated on plastic substrates by spin-coating [1–4]. Recently, the photolithography-based fabrication and direct transfer process of semiconducting nanowires (NWs) from a wafer has begun to emerge as a promising technology for the fabrication of such inorganic NWs on plastic substrates [5–7]. In actual practice, field-effect transistors (FETs) with channels composed of Si NWs transferred onto plastic substrates have revealed their usefulness as a functional component for flexible electronics [8–10]. Nevertheless, as yet the transferred NWs have not been used as a material for plastic optoelectronics. Therefore, in this study, we investigate the possibility of using the transferred Si NWs as an active component for optoelectronics on plastic substrates.

There have been several attempts to enhance the photocurrent efficiency of photodetectors by introducing new architectures based on transistors [11,12]. The established concept of the monolithic integrated system constructed with optoelectronic and electronic devices has prompted the idea that the transferred Si NWs can be used as the building blocks for a monolithic integrated system, because their length is sufficient to fabricate two functional devices on them. Thus, in this study, we propose a laterally coupled device composed of a *pn* heterojunction photodiode and a FET with a channel made of *p*-type Si NWs transferred onto a plastic substrate. Herein, the photodiode consists of the transferred *p*-type Si NWs and a sputtered *n*-type ZnO film. The transferred Si NWs in the laterally coupled device act not only as the channel for the FET, but also as the *p*-type material for the *pn* heterojunction photodiode. Figure 1 presents the electronic circuit diagram (a), the cross-sectional schematic of our laterally coupled device (b) and schematic energy band diagram of the heterojunction under thermal equilibrium (c). The energy gaps (E_G) of the ZnO film and the Si NWs are 3.4 and 1.1 eV, respectively, and the electron affinities of the ZnO film and the Si NWs are 4.19 and 4.05 eV, respectively. A *pn* heterojunction photodiode composed of the *p*-type Si NWs and the *n*-

type ZnO film is type-II. In this study, we examine the individual characteristics of the two-unit devices in the monolithic integrated system, and then investigate the performance of the photodiode modulated by operating the FET.

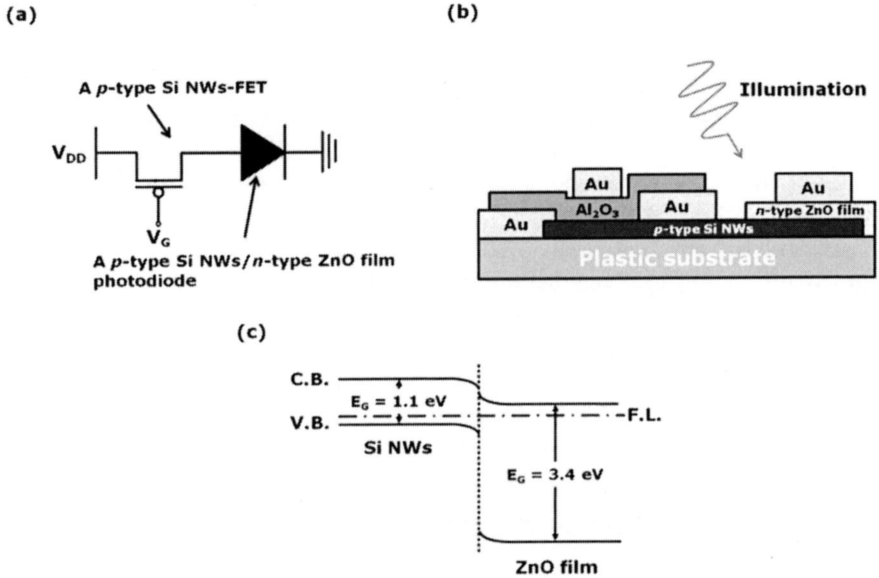

Figure 1. **(a)** Electronic circuit diagram; **(b)** cross-sectional schematic; and **(c)** schematic energy band diagram of a laterally coupled device composed of a *p*-type Si NWs-FET and a *p*-type Si NWs/*n*-type ZnO film photodiode.

2. EXPERIMENTAL PROCEDURE

P-type Si NWs were obtained from a boron doped Si bulk wafer with a doping concentration of $\sim 10^{17}$ cm^{-3} using photolithography and an anisotropic wet etching processes [13]. The Si NWs with a diameter of 200 nm and a length of 300 μm were directly transferred onto a plastic substrate. The laterally coupled device was fabricated using the following procedures: (i) the pattern for a ZnO film was formed on one side of the transferred Si NWs by a photolithography process; (ii) the ZnO film was deposited on the pre-patterned region by rf magnetron sputtering [14] (note that the width, length, and thickness of the deposited ZnO film were 1 mm, 0.8 mm, and 180 nm, respectively, and that the sputtered ZnO film under study had clear *n*-type characteristics and good crystallinity (Figure S1, ESI†); (iii) Au electrodes with a thickness of 100 nm were deposited by thermal evaporation for use as the source and drain electrodes of the FET and the negative electrode of the photodiode; (iv) an Al_2O_3 layer with a thickness of 20 nm was formed by atomic layer deposition for use as the gate insulating layer of the FET; (v) finally an Au electrode was formed as the gate electrode of the FET. Figure 2 shows the optical top-view image of the laterally coupled device constructed with the photodiode and the FET. The electrical characteristics of the two single devices and the laterally coupled device were measured with a semiconductor parameter analyzer (Agilent 4155C) in air at room temperature. The excitation source for the photocurrent measurement was the 325-nm wavelength light from a He-Cd laser with a power density of 48 W/cm^2 and neutral density filters were used to adjust the power of the incident light. The 325-nm wavelength light was irradiated onto the top-view side of the device. The illumination area was 28.3 mm^2, and the entire area of the device was illuminated by the light since the beam size of the 325-nm wavelength light was much larger than the device size (8 mm^2).

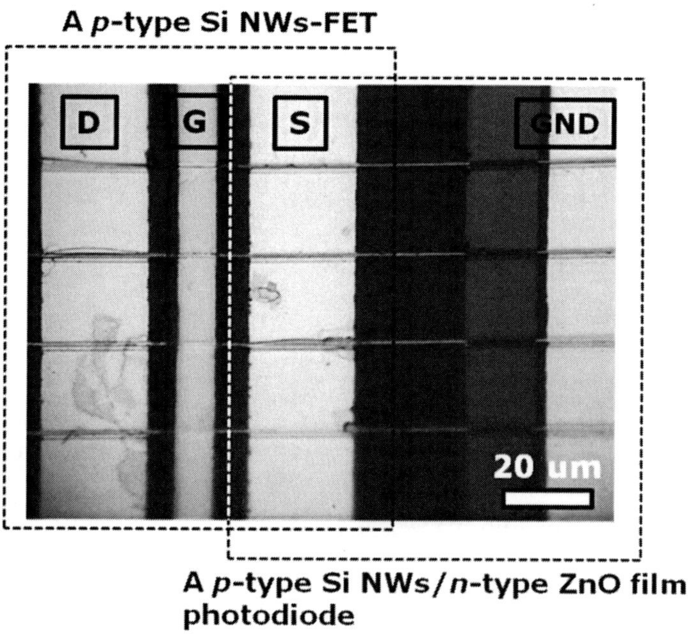

Figure 2. The optical top-view image of the laterally coupled device.

3. RESULTS AND DISCUSSION

3.1. The Optoelectronic Characteristics of the Single Photodiode Composed of the Transferred p-type Si NWs and the n-type ZnO Film

Figure 3(a) shows the current-voltage (I–V) curves for the dark current and photocurrent for the single photodiode. The line shape of the I–V characteristics indicates the formation of a *pn* heterojunction between the transferred *p*-type Si NWs and the *n*-type ZnO film. The rectification ratio estimated at bias voltages of ±2.5 V increases from 2.3 in the dark to 5.5 under the illumination of light with a power density of 48 mW/cm^2. The inset of Figure 3(a) shows the magnitude of the photocurrent as a function of the power density of the incident light at a bias voltage of 2.5 V. The magnitude of the photocurrent is proportional to the power density of the incident light. Note that the power density of the incident light increases from 8 to 48 mW/cm^2 in intervals of 8 mW/cm^2. When light with a power density of 48 mW/cm^2 is illuminated onto the photodiode at a bias voltage of 2.5 V, the photocurrent efficiency and on/off ratio are determined to be about 2.3 μA/W and 14.4, respectively.

Figure 3(b) shows the photocurrent transient with a 50-second interval at a bias voltage of 2.5 V under the illumination of the 325-nm wavelength light with a power density of 48 mW/cm^2. The photocurrent transient characteristics of our photodiode are similar to those reported in the studies of other research groups on *pn* heterojunction photodiodes composed of *n*-type ZnO and *p*-type Si [15,16], except for the phenomenon wherein the magnitude of the dark current grows slightly after the light is turned off. This phenomenon is attributed to the long length of our photodiode (The gap between the positive and negative electrode is estimated to be 40 μm). The photogenerated charge carriers adjacent to the electrodes rapidly move to the electrodes as soon as the light is turned off, resulting in initial fast photocurrent decay. On the other hand, the trapped charge carriers in the localized sites far away from electrodes are detrapped, and then drift toward the electrodes in the internal electric field. Consequently, this transport of the charge carriers is responsible for the increase of the dark current after the light is turned off. Figure 3(c) shows schematic energy band diagram of a *p*-type Si NWs/*n*-type ZnO film heterojunction photodiode in forward bias condition under the illumination of the 325-nm wavelength light. The difference (140 meV) between the electron affinities of the ZnO film and the Si NWs is much larger than the binding energy (60 meV) of excitons formed in the ZnO film. The large difference of the electron affinity enables the breaking of the excitons, and it can be the strong driving force for the charge separation. The photocurrent mechanism is briefly described, as follows; the detailed description on the forming and breaking of excitons and the charge separation are omitted here, for simplicity. The charge carriers are photogenerated in both regions of the Si NWs and the ZnO film when the light is irradiated on the device. Holes photogenerated in the Si NWs cannot pass through the junction potential barrier corresponding to the valence band offset, whereas electrons photogenerated in ZnO film transport toward the opposite site (or toward the positive electrode) in the internal electric field. Simultaneously, the photogenerated electrons in the Si NWs and holes in the ZnO film transport to positive and negative electrode, respectively. Consequently, the photogenerated charge carriers contribute to the photocurrent of the photodiode.

3.2. The Electrical Characteristics of the Transferred p-type Si NWs-FET

The drain-source current-voltage (I_{DS}-V_{DS}) curves taken from the *p*-type Si NWs-FET connected electrically to the photodiode in the laterally coupled device are illustrated in Figure 4(a) for gate voltage (V_G) values ranging from +5 to −5 V with a step voltage of −1 V. The magnitude of I_{DS} increases as V_G decreases, indicating the *p*-type characteristics of the transferred Si NWs. Thus, this FET is able to be utilized as a switching component in the laterally coupled device. The transfer characteristics (left axis) and the transconductance (g_m) curve (right axis) of the *p*-type Si NWs-FET at a V_{DS} of 1 V are illustrated in Figure 4(b). In the transfer characteristics, the I_{on}/I_{off} ratio taken from the Si NWs-FET is estimated to be 2.6×10^5, and the transconductance is found to be 617 nS at $V_G = 2$ V. The threshold voltage of the FET is evaluated from a linear fitting to the curve of the I_{DS} *versus* V_G with the extracted value of ~2.5 V. The field-effect mobility taken from the *p*-type Si NWs-FET is calculated to be 21.9 $cm^2/V\,s$ at a V_{DS} of 1 V.

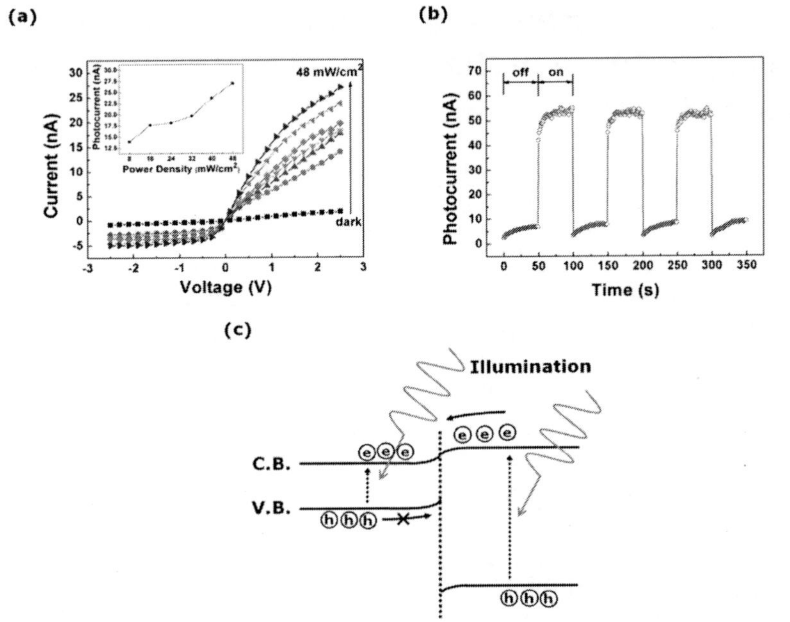

Figure 3. Optoelectronic characteristics of a *p*-type Si NW/*n*-type ZnO film photodiode.**(a)** The photocurrent characteristics under the illumination of the 325-nm wavelength light and the magnitude of the photocurrent depending on the power density of the light (inset); **(b)** the photocurrent transients under periodic illumination of light at a bias of 2.5 V; and **(c)** schematic energy band diagram in forward bias condition under the illumination of the 325-nm wavelength light.

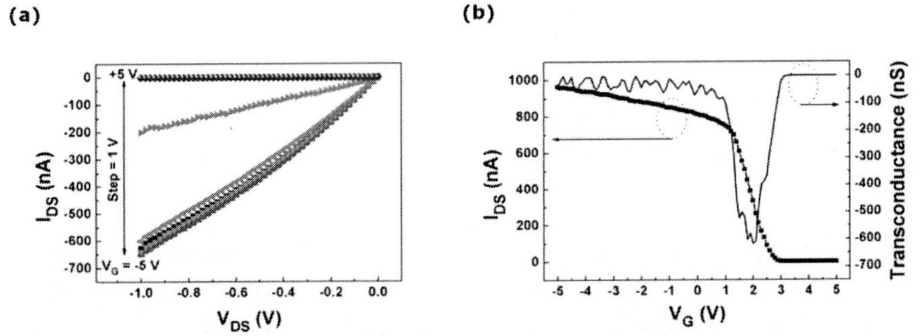

Figure 4. The electrical characteristics of the *p*-type Si NWs-FET. **(a)** The output characteristics and **(b)** the transfer characteristics (left axis) and the transconductance curve (right axis).

3.3. Characteristics of the Photodiode Modulated by the Operation of the FET in the Laterally Coupled Device

The drain electrode of the FET is biased at a voltage (V_{DD}) of 2.5 V in order to operate the laterally coupled device composed of the photodiode and the FET. Then, the characteristics of the photodiode are investigated for V_G values ranging from +3 to −9 V with a step voltage of −3 V. The photocurrent transients of the photodiode in the laterally coupled device are observed under the illumination of the 325-nm wavelength light with a 50-second interval as a function of V_G, as presented in Figure 5(a). The magnitude of the dark current and photocurrent increases simultaneously as the value of V_G decreases.Figure 5(b) shows the dependence of the photocurrent efficiency and the on/off ratio on the gate voltage applied to the FET in the laterally coupled device. The photocurrent efficiency remains nearly constant as the gate voltage deceases from +3 to −6 V. The photocurrent efficiency of the photodiode coupled with the FET increases to 10.0 μA/W when the FET at a V_G of −9 V is switched on. On the other hand, the on/off ratio of the photodiode gradually decreases from 20 to 4 when the gate voltage is changed from +3 to −9 V. The dramatic increase in the magnitude of the dark current measured at a gate voltage of −9 V is responsible for the low on/off ratio (Figure S2, ESI†).

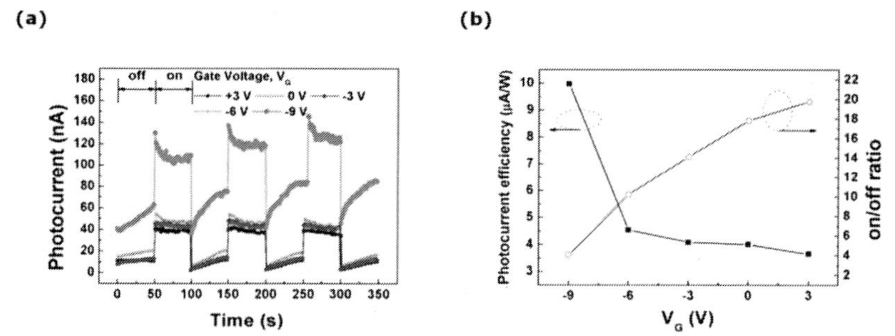

Figure 5. (a) The photocurrent transients of the photodiode coupled with the FET as a function of the gate voltage, and **(b)** the photocurrent efficiency and the on/off ratio as a function of the gate voltage.

In comparison with the photocurrent transient of the single photodiode (as shown in Figure 3(b)), some differences ascribed to the potential barrier in the *pn* heterojunction are found in the characteristics of the photocurrent and the dark current. When a positive gate voltage of +3 V is applied, the conduction and valance bands in the Si NWs are raised up, which enlarges the height of the potential barrier for the photogenerated holes as demonstrated in Figure 6(a). This high potential barrier prevents the transport of the photogenerated holes toward the opposite electrode at positive gate voltages. Meanwhile, when a gate voltage of −9 V is applied, the photocurrent and the dark current considerably increase, due to the lowering of the height of the potential barrier between the valance bands of the ZnO film and the Si NWs as exhibited inFigure 6(b). Thus, the photogenerated holes move readily toward the negative electrode of the photodiode. The photocurrent efficiency of the photodiode coupled with the FET increases remarkably when the FET at V_G of −9 V is sufficiently switched on.

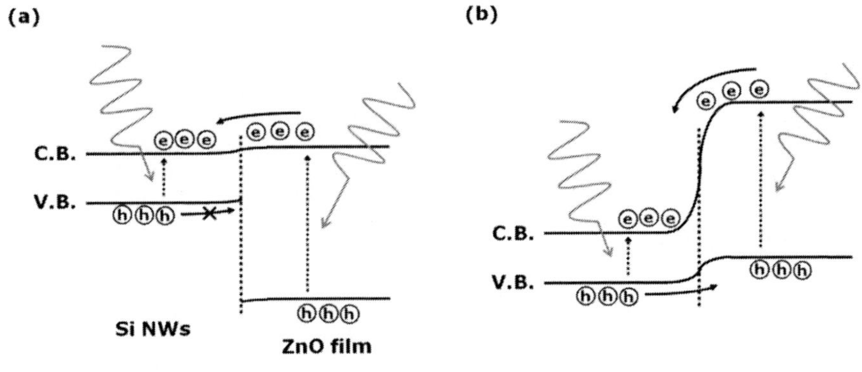

Figure 6. Schematic energy band diagrams of a *p*-type Si NWs/*n*-type ZnO film heterojunction photodiode coupled with a Si NWs-FET in forward bias condition under the illumination of the 325-nm wavelength light. The gate voltages of **(a)** +3 and**(b)** −9 V are applied on the gate electrode, respectively.

4. CONCLUSIONS

In this study, a laterally coupled device composed of a photodiode and a FET is fabricated on a plastic substrate by utilizing transferred Si NWs. The photodiode consists of transferred *p*-type Si NWs and a sputtered *n*-

type ZnO film. The photocurrent efficiency of the photodiode coupled with a FET increases remarkably when the FET is switched on at a V_G of −9 V, and the value of 10.0 μA/W is five times larger than that of a single photodiode without any FET. Our study demonstrates that the photocurrent efficiency of a photodiode can be improved in a monolithic integrated system composed of a photodiode and FET and that the transferred Si NWs are an appropriate inorganic nanomaterial for the fabrication of a monolithic integrated system on a plastic substrate.

ACKNOWLEDGMENTS

This work was partly supported by the IT R&D program of MKE/KEIT (10030559, Development of next generation high performance organic/nano materials and printing process technology), the Nano R&D Program (M10703000980-08M0300-98010), and World Class University (WCU, R32-2008-000-10082-0) Project of the Ministry of Education, Science and Technology (Korea Science and Engineering Foundation).

REFERENCES

1. Jang, J; Cho, K; Byun, K; Hong, WS; Kim, S. Optoelectronic Characteristics of HgSe Nanoparticle Film Spin-Coated on Flexible Plastic Substrates. *Jpn J Appl Phys* **2010**, *49*, 030210:1–030210:3.

2. Byun, K; Cho, K; Kim, S. Sintering Effect on the Optoelectronic Characteristics of HgSe Nanoparticle Film on Plastic Substrates. *Mater. Chem. Phys* **2010**, *122*, 246–249.

3. Huang, J; Li, G; Yang, Y. A Semi-Transparent Plastic Solar Cell Fabricated by a Lamination Process. *Adv. Mater* **2008**,*20*, 415–419.

4. Lau, SP; Yang, H; Yu, SF; Yeun, C; Leong, ESP; Li, H; Hng, HH. Flexible Ultraviolet Random Lasers Based on Nanoparticles. *Small* **2005**, *1*, 956–959.

5. Sun, Y; Kim, HS; Menard, E; Kim, S; Adesida, I; John, AR. Printed Arrays of Aligend GaAs Wires for Flexible Transistors, Diode, and Circuits on Plastic Substrates. *Small* **2006**, *11*, 1330–1334.

6. Michael, CM; Habib, A; Dunwei, W; James, RH. Highly Ordered Nanowire Arrays on Plastic Substrates for Ultrasensitive Flexible Chemical Sensors. *Nat. Mater* **2007**, *6*, 379–384.

7. Sun, Y; John, AR. Fabricating Semiconductor Nano/Microwires and Transfer Printing Ordered Arrays of Them onto Plastic Substrates. *Nano Lett* **2004**, *4*, 1953–1959.

8. Lee, KN; Jung, SW; Kim, WH; Lee, MH; Shin, KS; Seong, WK. Well Controlled Assembly of Silicon Nanowires by Nanowire Transfer Methode. *Nanotechnol* **2007**, *18*, 445302:1–445302:7.

9. Chung, EA; Koo, J; Lee, M; Jeong, DY; Kim, S. Enhancement-Mode Silicon Nanowire Field-Effect Transistors on Plastic Substrates. *Small* **2009**, *5*, 1821–1835.

10. Lee, M; Koo, J; Chung, EA; Jeong, DY; Koo, YS; Kim, S. Silicon Nanowire-Based Tunneling Field-Effect Transistors on Flexible Plastic Substrates. *Nanotechnol* **2009**, *20*, 455201.

11. Bae, HS; Im, S. Ultraviolet Detecting Properties of ZnO-Based Thin Film Transtors. *Thin Solid Films* **2004**, *469–470*, 75–79.

12. Bae, HS; Yoon, MH; Kim, JH; Im, S. Photodetecting Properties of ZnO-Based Thin-Film Transistors. *Appl. Phys. Lett***2003**, *83*, 5313–5315.

13. Koo, J; Kim, S. Charge Transport Modulation of Silicon Nanowire by O_2 Plasma. *Solid State Sci* **2009**, *11*, 1870–1874.

14. Masuda, S; Kitamura, K; Okunura, Y; Miyatake, S; Tabata, H; Kawai, T. Transparent Thin Film Transistors Using ZnO as an Active Channel Layer and Their Electrical Properties. *J. Appl. Phys* **2003**, *93*, 1624–1630.

15. Mridha, S; Basak, D. Ultraviolet and Visible Photoresponse Properties of n-ZnO/p-Si Heterojunction. *J Appl Phys* **2007**,*101*, 083102:1–083102:5.

16. Ghosh, R; Basak, D. Electrical and Ultraviolet Photoresponse Properties of Quasialigned ZnO Nanowires/p-Si Heterojunction. *App Phys Lett* 2007, *90*, 243106:1–243106:3.

11

PHOTODIODES AS OPTICAL RADIATION MEASUREMENT STANDARDS

Ana Luz Muñoz Zurita[1], Joaquín Campos Acosta[2], Alejandro Ferrero Turrión[2] and Alicia Pons Aglio[2]

[1] Universidad Autónoma de Coahuila, Campus Torreón, Faculty of Enginering Mechanical and electrical. Torreón, Coahuila, México

[2] Consejo Superior de Investigaciones Científicas (CSIC), Instituto de óptica "Daza de Valdés", Madrid, España

1. INTRODUCTION

Photodiodes for optical radiation measurements are used without reverse bias in most applications since this operation yields the lowest dark current. To obtain photodiodes that operate at a low bias and have a low dark current, it is necessary to produce epitaxial layers that are pure and have few defects (such as dislocations, point defects, and impurity precipitates). Furthermore, a planar device structure requires that a guard ring be used to keep the electric field around the photoreceptive area from increasing too much. Fabrication and processing technologies such as impurity diffusion, ion implantation, and passivation play important roles in the production of reliable photodetectors.

From a radiometric point of view, the photodetectors important characteristics are: Speed of response (characterized by the bandwidth of the frequency response or the Full Width Half Maximum (FWHM) of the pulse response), responsivity (determined as the ratio of current out the detector to the incident optical power on the device), sensitivity (defined as the minimal input power that can still be detected which, as a first approximation, is defined as the optical power which generates an electrical signal equal to that due to noise of the diode) and response linearity. These quantities defined the basic radiometrical behavior of any detector. For those detectors having large area, as it may be the case for some photodiodes, knowing the response uniformity of the sensitive area is important too, especially when the incident beam diameter is much smaller than the detector sensitive surface. A high nonuniformity would produce measurement errors when the detector is used at different positions, errors that have to be taken into account for the final accuracy of the measurement.

To determine those radiometric features in photodiodes and learn how they change with wavelength, for instance, it is a good approach to start by analyzing. The physical phenomena involved in the detection. When light impinges on a detector, various physical processes occur; part of the incident light is reflected at the sensitive surface, while the rest passes inside the detector, where can be partially, because of losses due to absorption, converted into an electronic signal. Then the photodetector response is conditioned by the amount of absorbed light, but for evaluating the incident power one has to know the ratios of the reflected, absorbed, and converted power as well. Taking into account these phenomena, the short circuit response of a photodiode can be written as

$$I(\lambda)=I_0+(1-\rho(\lambda))\varepsilon(\lambda)\frac{\lambda}{k}\varnothing(\lambda) \tag{1}$$

Where *I 0* is the dark response, ρ(λ) is the photodiode's reflectance, λ is the radiation wave length, ε (λ) is the photodiode's internal quantum efficiency,*k* is a constant that takes into account other fundamental physical constants and ϕ (λ) is the spectral radiant flux incident on the photodiode. According to this equation, the incident radiant flux can be determined from measuring the photodiode's response as far as its spectral reflectance and internal quantum efficiency are known. Then photodiodes are good devices for radiant flux standards.

Silicon and InPphotodiodes from different manufacturers have got rather low noise level, good response uniformity over the sensitive surface and a wide dynamic range. Therefore they are good devices to build radiometers in the visible and NIR spectral region in many different applications, particularly for building up spectroradiometric scales for radiant flux measurements.

Back to equation (1), if photodiode's reflectance and internal quantum efficiency were known, the photodiode's responsivity would be known without being compared to another standard radiometer; i. e. the photodiode would be an absolute standard for optical radiation measurements [1, 2, 3].

This idea was firstly developed for silicon photodiodes in the eighties, once the technology was able to produce low defects photodiodes [4]. Following this reference, the reflectance could be approached from a superimposed thin layers model. By knowing the thicknesses of the layers and the optical constants of the materials, it is possible to determine the device reflectance. However, this information is not completely available for InP photodiodes: the actual thickness of the layers is not known and optical constants of materials are only approximately known for bulk. Nevertheless it's possible to measure reflectance at some wavelengths and to fit the thicknesses of a layer model that would reproduce those experimental values.

The internal quantum efficiency cannot be determined as for Si. Since InP photodiodes are hetero-junctions rather than homo-junctions as silicon photodiodes are. In the other hand, since the internal structure is not accurately known, it is not possible to model the internal quantum efficiency without having experimental values for it.

Therefore the attainable scope at present is just to obtain a model to be able to calculate spectral responsivity values at any wavelength. To get this, a model has been developed to calculate reflectance values from experimental ones at some wavelengths and another model has been developed to interpolate spectral internal quantum efficiency values from some values got from reflectance and responsivity measurements at some wavelengths. Both models will be presented in this chapter.

2. SPECTRAL RESPONSIVITY SCALE IN THE VISIBLE RANGE BASED ON SINGLE SILICON PHOTODIODES.

A spectral responsivity scale means that the responsivity is known at every wavelength within the response range of interest and it would be desirable to know it for all the other parameters associated with a beam: angle of incidence, divergence or polarization.

Aspectral responsivity scale in the visible range can be created by calibrating a silicon trap detector at several laser wavelengths against ahigh accuracy primary standard such as an electrically calibrated cryogenic radiometer. This method provides a very certain value for the responsivity at specific wavelengths as those of lasers (for instance 406.7 nm, 441.3 nm, 488.0 nm, 514.5 nm, 568.2 nm, 647.1 nm and 676.4 nm). From there single elements detectors, most suitable for some applications, can be calibrated against that trap detector at those wavelengths to define the working scale.

The spectral responsivity of silicon photodiodes is given by the well-known equation

$$R(\lambda)=(1-\rho(\lambda))\varepsilon(\lambda)\frac{\lambda}{k} \tag{2}$$

This chapter describes the results obtained for the responsivity of the photodiodes by using a model to calculate the diode's reflectance from experimental measurements and a model for the internal quantum efficiency, which is also fitted to experimental values. Based on the models, the fitting errors and the uncertainty of reflectance and responsivity measurements, the uncertainty of the responsivity scale is calculated according to the ISO recommendations.

3. REFLECTANCE EVALUATION OF SILICON PHOTODIODES

From the reflectance point of view, a silicon photodiode can be considered as a system formed by a flat transparent film over an absorbing medium. The flat film is the silicon oxide and the absorbing medium is the silicon substrate. The reflectance of such a system is given by [5]

$$\rho=\frac{\left[r_{12}^2+\rho_{23}^2+2r_{12}\rho_{23}\cos(\phi_{23}+2\beta)\right]}{1+r_{12}^2\rho_{23}^2+2r_{12}\rho_{23}\cos(\phi_{23}+2\beta)} \tag{3}$$

where $r12$ is the amplitude of the reflection coefficient from air to silicon oxide, $\rho 23$ is the amplitude of the reflection coefficient from silicon oxide to silicon, $\varphi 23$ is the phase change at the interface silicon oxide–silicon and $\beta = 2\pi n2h\cos(\theta 2)/\lambda 0$, with h the thickness of SiO2, $n2$

the refractive index of SiO2 and $\theta 2$ the refraction angle at the air–oxide interface. These variables change with the angle of incidence and the light polarization, so the reflectance value will be known if the silicon oxide thickness, the angle of incidence, the refractive index and the light polarization status are known. This reflectance model has been already tested for another type of silicon photodiode from the same manufacturer [6].

Spectral values of the refractive index are available in the literature. In this work values have been obtained from those given in [7]. The index of refraction of silicon oxide has been interpolated by fitting a polynomial to data; the real part of the refractive index of silicon has been obtained by fitting a polynomial in $1/\lambda$ and the imaginary part by fitting an exponential decay in λ.Reflectance was measured with an angle of incidence of 4 in our reference spectrophotometer, using p-polarized light, at the laser wavelengths for which the diodes were calibrated against the trap: 406.7 nm, 441.3 nm, 488.0 nm, 514.5 nm, 568.2 nm, 647.1 nm and 676.4 nm. By fitting equation 3 to measurement results, the silicon oxide thickness was obtained for every photodiode, as shown in table 1. The fitting error in this table is the parameter given by the fitting software.

Table 1. Silicon oxide thickness fitted to reflectance measurements

Photodiode	SiO2 thickness/nm	Fitting error/nm
CIRI	29.58	0.19
SiN	28.84	0.17
Si1	29.93	0.19

The fitting is very good for wavelengths longer than 500 nm, getting worse for shorter wavelengths, as can be seen in figure 1 for one of the photodiodes studied. The same results are obtained for the three photodiodes studied in this work.

This agrees also with [2]. Probably it is due to the measurement bandwidth. For convenience, the reflectance was measured in our reference spectrophotometer with a bandwidth of 5 nm in order to have a good signal-to-noise ratio at the shortest wavelengths. But in this region the first and second derivatives of reflectance are higher than in the middle visible, so the increased bandwidth produces an effective reflectance value that differs significantly from the spectral value. For this reason, reflectance values below 500 nm were not used in the final fitting process to obtain the thickness.

Using thickness values given in table 1, the reflectance of the photodiodes at normal incidence can be calculated, and from them and the responsivity values measured against the trap detector, the photodiodes' internal quantum efficiency can be calculated according to (2). Using a model based in physical laws rather than experimental equations allows obtaining the physical quantity for different circumstances, such as different angles of incidence, for instance.

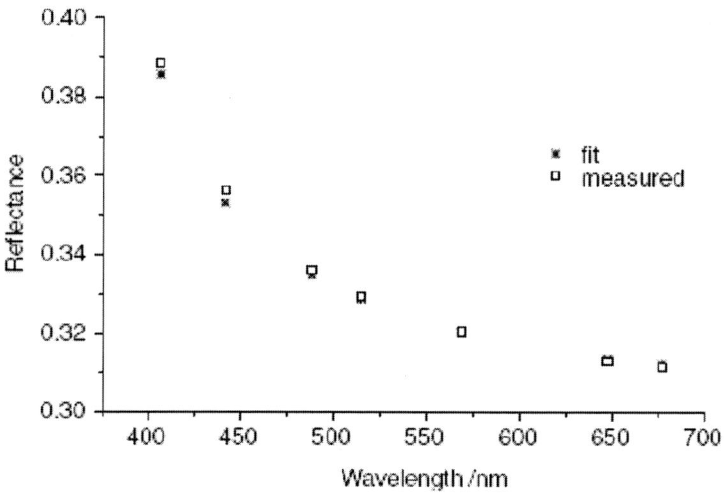

Figure 1. Spectral reflectance values of photodiode CIRI and fitted values according to **equation (3)**.

5. INTERNAL QUANTUM EFFICIENCY OF SILICON PHOTODIODES

To spectrally know the internal quantum efficiency we have used the model developed by Gentile *et al*[2], based on that from Geist and Baltes [14] and improved by Werner *et al* [15]. The internal quantum efficiency is given by

$$\varepsilon(\lambda)=P_f+\frac{1-P_f}{\alpha(\lambda)T}\{1-\exp[-T\alpha(\lambda)]\}-\frac{1-P_b}{(D-T)\alpha(\lambda)}\{\exp[-T\alpha(\lambda)]-\exp[D\alpha(\lambda)]\}-P_b\exp\exp[h\alpha(\lambda)]+R_{back}\exp[h\alpha(\lambda)]P_b \quad (4)$$

where *Pf* is the collection efficiency at the front, *T* is the junction depth, *Pb* is the collection efficiency at the silicon bulk region, which starts at depth *D*, *h* is the photodiode's length, *R*back is the reflectance at the photodiode's back surface and α is the absorption coefficient.According to Gentile *et al* [2] a simplified model can be used if the model is to be applied to wavelengths shorter than 920 nm. This model is obtained from the previous equation by deleting the last two terms. Then, the quantum efficiency can be obtained from

$$\varepsilon(\lambda)=\mathrm{Pf}+(1-\mathrm{Pf})/\alpha(\lambda)\mathrm{T}\,\{1-\exp[-\mathrm{T}\alpha(\lambda)]\}-(1-\mathrm{Pb})/((\mathrm{D}-\mathrm{T})\alpha(\lambda))\,\{\exp[-\mathrm{T}\alpha(\lambda)]-\exp[\mathrm{D}\alpha(\lambda)]\} \quad (5)$$

This model has been fitted to the calculated internal quantum efficiency values by a non-linear squared method.

The parameters' initial values were taken from Gentile *et al* [2]. The goodness of the fit can be seen infigure 2, where values for one of the studied photodiodes are shown. The same results are obtained for the three photodiodes studied in this work. The main difference between the fitted values of the internal quantum efficiency and those calculated from the responsivity and reflectance measurements is about 10^{-3}, which agrees well with results given by other authors, e.g.[2,9].

Another point that can be discussed is how far the internal quantum efficiency can be extrapolated. Using this simplified model and fitting with values corresponding to wavelengths shorter than 700 nm, quantum efficiency values continue to increase very slightly to 900 nm at least. This is not what really happens in the photodiode, so there will be an upper limit for the extrapolation. This limit will depend on the uncertainty allowable to the responsivity value and will be discussed in the following section.

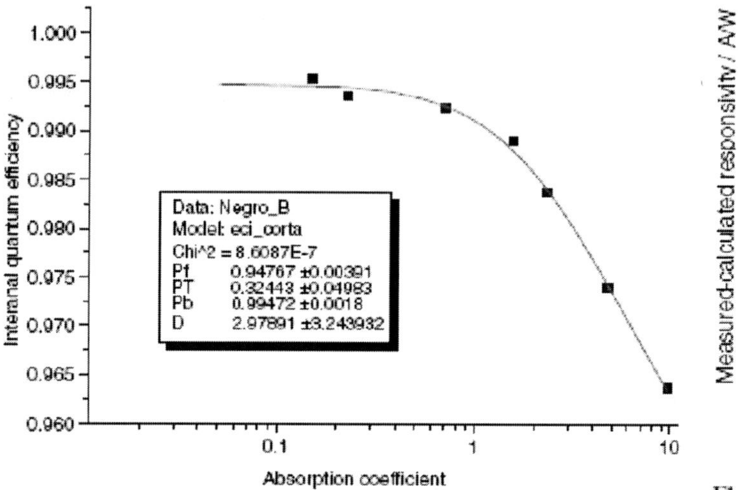

Fi

Figure 2. Experimental internal quantum efficiency values ofphotodiode SiN and fitted values according to **equation (5)** againstthe absorption coefficient.

6. SPECTRAL RESPONSIVITY VALUES OF SILICON PHOTODIODES

Responsivity of detectors has been calculated with the model described previously and the parameters obtained by the fitting process by using (2), (3) and (5). The agreement between the calculated values and those measured against the trap is excellent as can be seen in figure 3 for one of the photodiodes studied. This result is just a check of the consistency of the method. Nevertheless, it can be seen that most calculated values are smaller than the measured ones. This might be due to the independent fitting of reflectance and quantum efficiency values and their functional forms, but it may also be due to the presence of a systematic error in the measurements. Some research will have to be done in the future to clarify this.

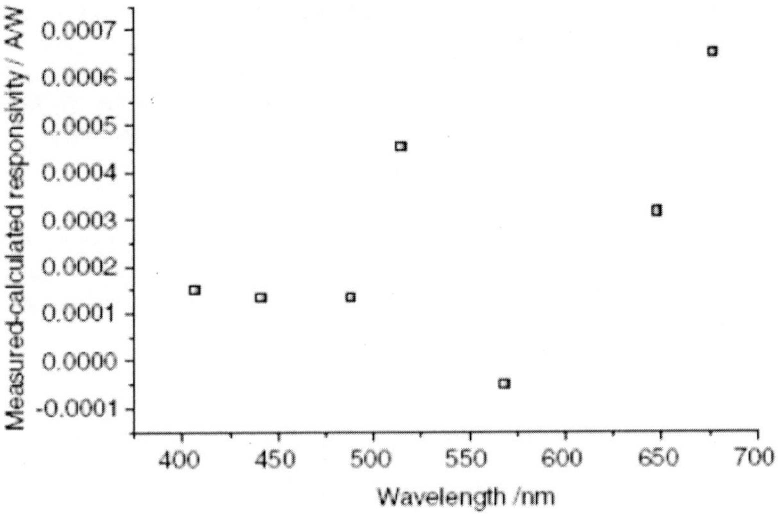

Figure 3. Spectral Difference between calculated and measured spectral responsivity values for photodiode CIRI as a function of wavelength.

7. SPECTRAL RESPONSIVITY SCALE IN THE NEAR IR RANGE BASED ON SINGLE INP/INGAAS PHOTODIODES

As in the visible range, semiconductor photodiodes are the best choice for establishing spectral responsivity scales in the near IR range. The first attempt was to use germanium photodiodes, since its gap allowed to obtain a device responding to wavelengths lower than 1.6 μm, approximately, depending on temperature. However germanium photodiodes have got a rather high dark current and lower shunt resistance than silicon, then they are not so useful for optical radiation detection. Since optical communications were demanding better detectors to enlarge their use, other photodiodes were developed in this spectral region of great interest. Since no other single element semiconductor was possible, semiconductor hetero-junctions were developed. A hetero-junction is a junction formed between two semiconductors with different band-gaps. Of course building such devices is not straightforward since the lattice parameters have to be matched, but this is not the subject of this chapter and many good references may be found in literature [17].

The group known as III-V hetero-structures has yield different photodiodes in the near IR range, particularly those based on InP/InGaAs has yield very good devices for the spectral range covered by germanium photodiodes. This hetero-structure has got two junctions in fact. The InGaAs material, having a lower gap, is kept in between two layers on InPwhose gap is bigger and hence transparent to the wavelength region used in optical communications: the nondispersion wavelength (1.3μm) and the loss minimum wavelength (1.55μm). The radiometric characteristics of these InP-based photodetectors are superior to those of conventional photodiodes composed of elemental Germanium. Because of that they have replaced germanium in almost every application.

By using a hetero-structure, which hadn't been used in group IV elemental semiconductors such as Si and Ge, new concepts and new designs for high performance photodetectors have been developed.For example, the absorption region for a specific spectral range can be confined to a limited inner layer, avoiding typical high recombination rates of charge carriers at the first interfaceof the photodiode and getting a higher internal quantum efficiency.

Recently InGaAs/InP avalanche photodiodes (APDs) with a SAM (separation of absorption and multiplication) configuration have become commercially available. The SAM configuration is thought to be necessary for high performance APDs utilizing long wavelengths.

InGaAs/InPphotodetectors are used for maintaining the scale of spectral responsivityup to 1.7 μm in many laboratories [17, 19].In addition they are exploited in instruments for measuring optical radiation within the near infrared (NIR) range (800 nm -1600 nm). From this point of view, these photodiodes are like other and their response is given by equations (1) and (2). Therefore to know their reflectance and internal quantum efficiency is the key for defining the spectral responsivity scale in this range.

Next experimental values for those properties measured in our laboratory for devices built by different manufacturers will be presented.

8. MEASUREMENT OF INP PHOTODIODE'S REFLECTANCE

To realize our experiments related to measuring the reflectance of InGaAs/InP photodiodes the experimental set-up presented in figure 4 hasbeen arranged.

An incandescence lamp is the white light source imaged at the input slit of the monochromator. This lamp was able to cover the spectral range from 800 nm to 1600 nm and appropriate blocking filters for second – order wavelengths were added to the monochromator. After the monochromator, a linear polarizer and a beam splitter, which serves to monitor temporal power fluctuations,were placed. A germanium photodiode was used as the monitoring reference photodetector. More details can be seen in reference [20].

The experimental set-up included an optical system of mirrors, which consists of two parts. An upper part (see mirror 7 and germanium photodiode 9) realized monitoring temporal fluctuations of light power. A bottom part (see mirrors 8, 11; InGaAs/InP-photodiode 10, and and germanium photodiode 12) formed an image of the monochromator's exit slit on the sensitive surfaces of photodiodes. The angle of incidence was equal to 7.4 º which was accepted as the normal incidence in this train of measurements.

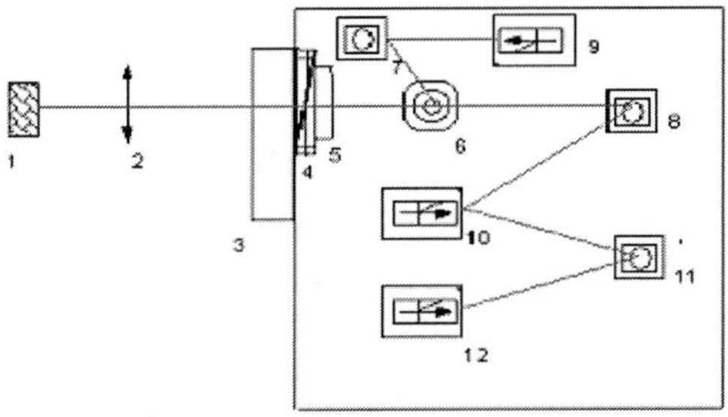

1. Source light (Incadescent lamp)
2. Lens
3. Monochromator
4. Polarizer
5.Stopper (Choper)
6. Beam splitter AR34
7,8,11. Mirrors
9,12. Germanium detectors
10. InGaAs/InP detectors with is exchanged with standart mirror BK47

Figure 4. Experimental set-up for measuring the reflectance InGaAs/InP photodiodes

The measurement method consists in comparing the response from a germanium photodiode to the radiation reflected by the InGaAs/InP photodiode with the response from an aluminum standard mirror whose reflectance is measured as in [21], so that [20]:

$$\rho(\lambda) = \frac{I_p(\lambda)}{I_m(\lambda)} \rho_m(\lambda) \tag{6}$$

Here,$I\,p\,(\lambda)$ is the response to the light reflected by

the InGaAs/InP, $I\ m\ (\lambda)$is the response to the light reflected by the mirror, and $\rho_m(\lambda)$ is the reflectance of a standard mirror. With this method the reflectance of photodiodes from different manufacturers hasbeen measured. One part of detectors had a round active area of 5 mm in diameter and the other part had a quadratic active area of 8 mm x 8 mm.

9. ANALYSIS OF REFLECTANCE OF INP PHOTODIODES

The polarization degree of light at the output the monochromator was different with varying the wavelength.The figures 5and 6 illustrate spectral dependences of the reflectance, which had been obtained from photodetectors belonging to three different manufacturers. Two types (photodiodes 1 and 4 and photodiodes 2 and 5) are 5 mm in diameter sensitive area and the third is an 8 mm in diameter sensitive area especially commercialized some years ago for developing spectral responsivity scales and no longer available in the market. Figure 5a, and 5b show that the reflectance of 5 mm in diameter detectors from both manufacturers has got a minimum in the region 1000 nm to 1600 nm, and they both are related to a structure of layers providing maximal responses in the spectral interval of mayor utility of these detectors in near IR:Optics communication [17]. The first photodiode, seeFigure 5a, whose reflectance was minimized, is more efficient that the second one, see figure 5b.

Reflectance in figure 6 is associated with a photodiode with rectangular active area. In this case the reflectance has two minima at 1000 nm and 1600 nm, but the reflectance has a maximum between these minima. This photodiode is older than previous ones, and it was produced by another manufacturer. One can remark that maybe it was produced without good enough control, because the structure of layers on the sensitive surface modifies the reflectance.

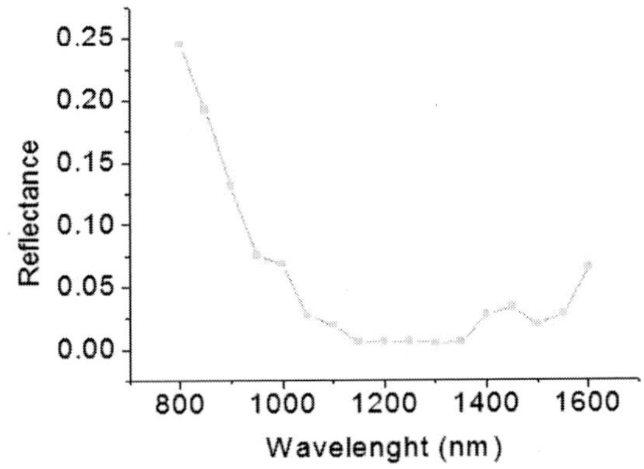

(a)

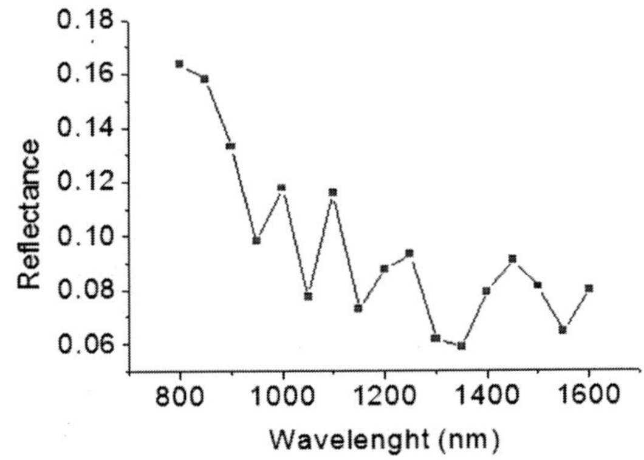

Figure 5. Detector with an active area 5 mmin diameter

The spectrum of reflectance for photodiodes 1 and 4, manufactured by the same company, is presents in figure 7. The reflectance was measured with linearly polarized and non-polarized lights, and these pair of measurements gives quite similar results. In fact, the difference was equal to approximately 2% for the angle of incidence used in this work. The same results are depicted for the photodiodes 2 and 5, manufactured by a second company. It is important that the results do not depend on the polarization state of the incident light when the angle of incidence is smaller than 10 degrees [22].

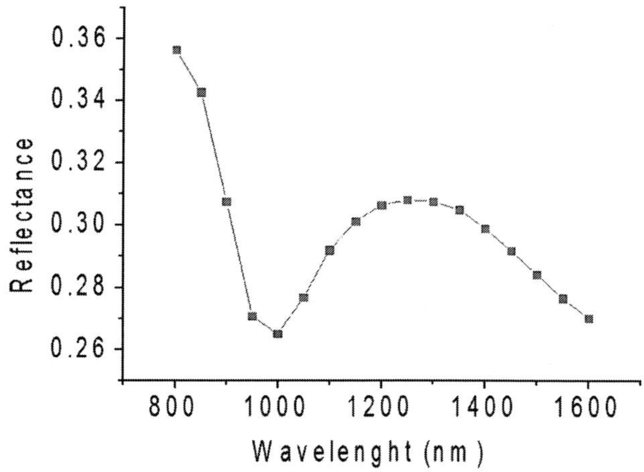

Figure 6. Detector with a rectangular aperture of 8 x 8mm

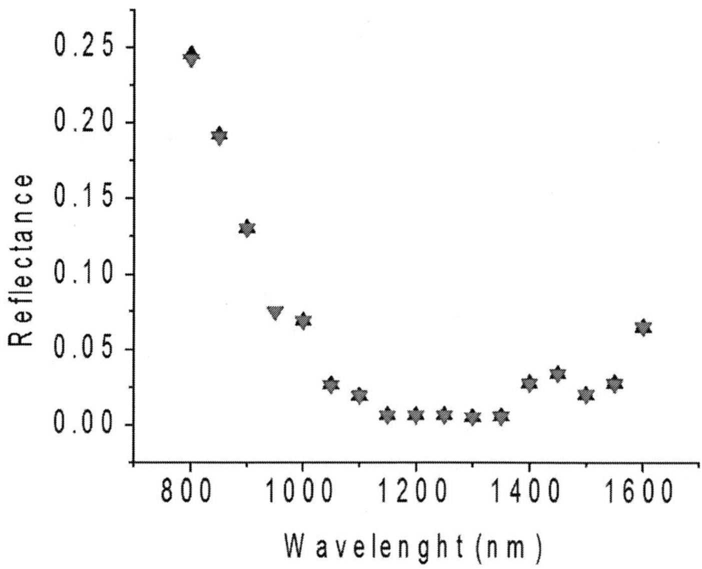

Figure 7. Spectrum of reflectance for photodiodes 1 and 4 from the same manufacturer.

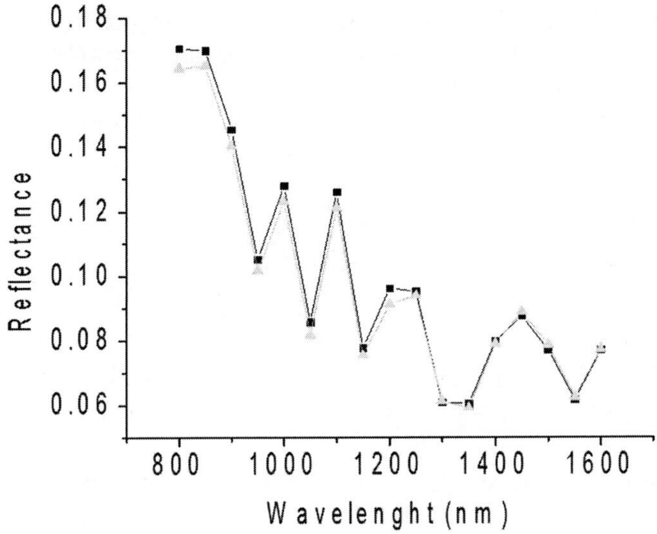

Figure 8. Spectrum of reflectance of photodiodes 2 and 5 from the same manufacturer.

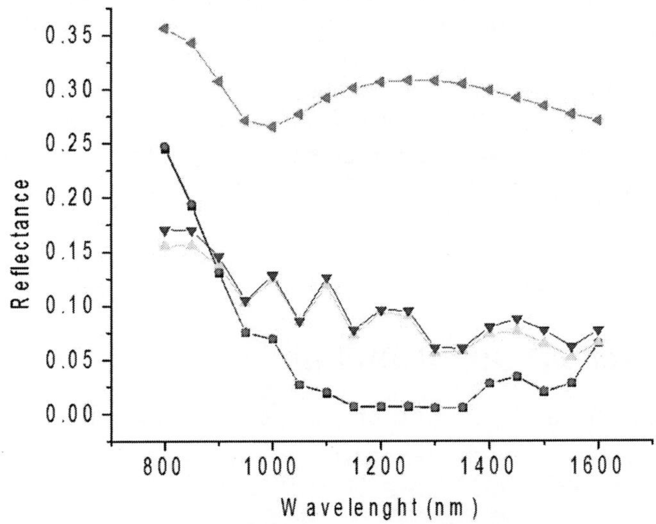

Figure 9. Comparison of reflectance of all photodiodes measured in this work.

All spectrums of reflectance are presented in Figure 9, with linearly polarized and non polarized light, so that it is possible to see the different behavior of the photodiodes in the near infrared wavelength. In fact, in this chapter we are studying the behavior of the photodetectors in the near infrared with the linearly polarized and non polarized light in the case of the polarized light the angle of incidence is smaller 10 angular degrees and is possible to observe the reflectance doesn't change its spectral behavior.

10. NEW QUANTUM INTERNAL EFFICIENCY MODEL OF SOME INPPHOTODETECTORS.

To determine the internal quantum efficiency of a photodiode it is necessary to know its responsivity (2). In this work, the responsivity, R(λ), was measured by direct comparison to an electrically calibrated pyroelectric radiometer (ECPR), obtaining responsivity values with an uncertainty of 1.2 % approximately, roughly the uncertainty of the ECPR. Spectral responsivity values of one photodiode from every manufacturer obtained from measurements are shown in figure 10 (analogous results are obtained for the other photodiode from the same manufacturer). From now on, the photodiodes will be identified as Ham, GPD and POL. Ham and GPD are photodiodes from different manufacturers and were identified before as photodiode 2 and photodiode 1, respectively. Both have got a 5 mm in diameter active area. Photodiode POL was identified before as photodiode 4 and has got an 8 mm side square active area. Figure 10 shows there is a noticeable difference in responsivity between them.

10.1. External Quantum Efficiency

It is obtained from the responsivity values according to the equation:

$$Q(\lambda) = \frac{R(\lambda)\,hc}{\lambda e} \tag{7}$$

Where *h*, *c* and *e* are the usual physical constants andλis the wavelength.Values obtained are presented in figure 11 for the same detectors as before. It can be clearly seen that the oldest detector (identified as POL) presents a lower external quantum efficiency than the other and that detector GPD presents a higher external quantum efficiency than detector HAM, which starts to decrease its quantum efficiency at a shorter wavelength. However, detector POL decreases less its quantum efficiency at wavelengths lower than the corresponding to the InGaAs gap. Perhaps this is mainly due to the tailoring of the hetero-structure done by the manufacturer. Detector POL was developed for realizing spectral responsivity scales, while the other two were developed for a better performance in the optical communications spectral range.

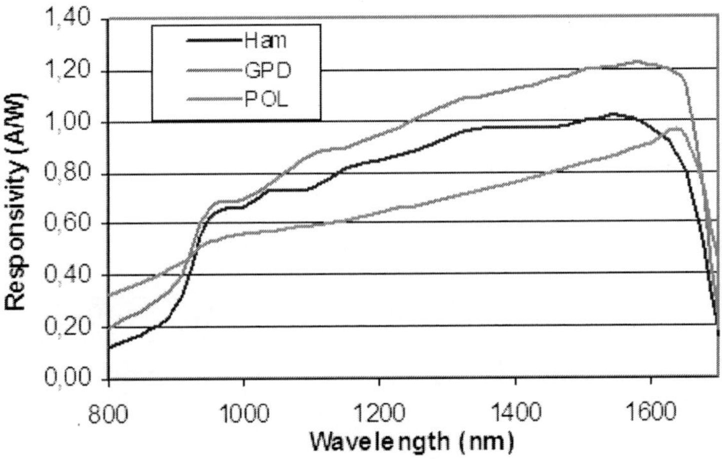

Figure 10. Spectral responsivity values of InPphotodiodes.

10.2. Internal Quantum Efficiency

Internal quantum efficiency is obtained from responsivity and reflectance by using (2). However those quantities have been measured at some wavelengths only, then it is necessary to develop a model to interpolate them at every wavelength within the response range. To develop such a model it is necessary to know the internal structure of the photodiode, as it was done for the silicon photodiode, but a enough precise structure is not available in the open literature. Since a structure has to be assumed to develop the model, the simplest one from literature has been adopted in this work (Fig. 12). It is more than likely that detector POL has particularly got a different structure.

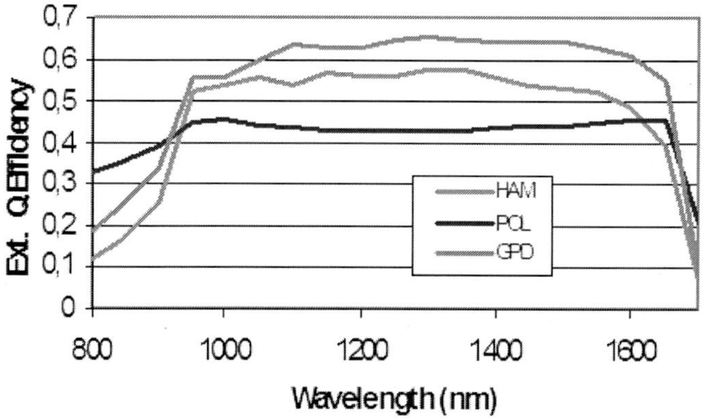

Figure 11. Spectral external quantum efficiency obtained from responsivity values

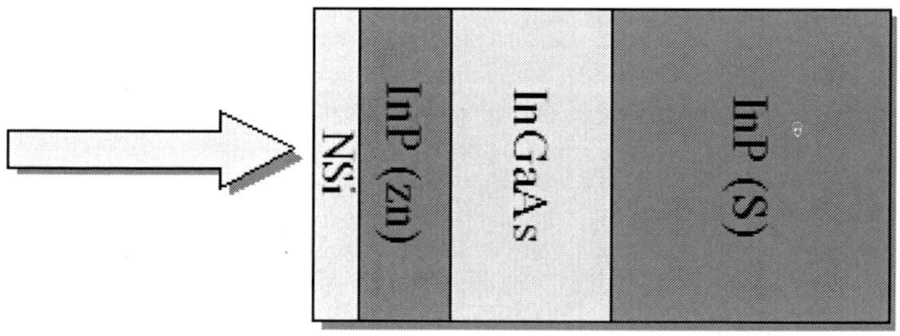

Figure 12. Internal Structure used in this work to model internal quantum efficiency of InP photodiodes

The first layer made on NSi is transparent in the wavelength range considered in this work. Probably it is placed in the photodiode as a passivation layer. Its thickness may be tailored by the manufacturer to spectrally adjust the device's reflectance.

Considering a structure as shown before (Fig. 12) and a simple model for the collection efficiency of carriers in every region given by a constant value: P_f,lower than 1 in the first region, 1 in the depletion region (mainly InGaAs) and P_b in the back region, and an "infinite" thickness for the diode, ε (λ) can be calculated by [23]:

$$\varepsilon(\lambda)=P_f(1-\exp(-\alpha T))+\exp(-\alpha T)-\exp(-\alpha T')+\exp(-\alpha' T') \\ -\exp(-\alpha' D')-\exp(-\alpha D')+\exp(1-P_b)\exp(-\alpha D) \quad (8)$$

Where T is the thickness at which collection efficiency becomes 1, T' is the thickness at which InGaAs region starts, D' is the the thickness at which the InP (S) starts and D is the thickness at which depletion region ends. By fitting this model to internal quantum efficiency values, the following parameters are obtained for every photodiode [23].

Table 2. Parameters fitting the model to experimental internal quantum efficiency

Photodiode	P_f	T	T'	D'	D	P_b
HAM	0	0.44	2.19	2.19	11.96	0.844
GPD	0	0.32	1.65	1.62	4351.16	0.960

Internal quantum efficiency values calculated from responsivity and reflectance (dots) and adjusted values following (8) are shown in **figures1 3** and **14** for photodiodes HAM and GPD, respectively. It can be seen that photodiode GPD has got an internal quantum efficiency very close to unity in the region from 1 μm to 1.6 μm, approximately. Both photodiodes have got internal quantum efficiency in this region nearly independent of wavelength. These two results are very important in order to try to develop an absolute radiometer based on InP photodiodes in the future.

The model does not fit well in the short wavelength region. Possibly this is because the structure of the detector is actually more complex or, perhaps, refraction index are not accurately known.

11. INTERPOLATION OF SPECTRAL REFLECTANCE

Finally, to have the spectral responsivity scale, it is necessary to interpolate spectral reflectance at any wavelength, what can be done by using a multilayer model [11]. Complex refraction index of the materials and thickness of the layers have to be known for interpolation. Refraction index has been obtained from [13] and other sources cited in there and are shown in figure 15.

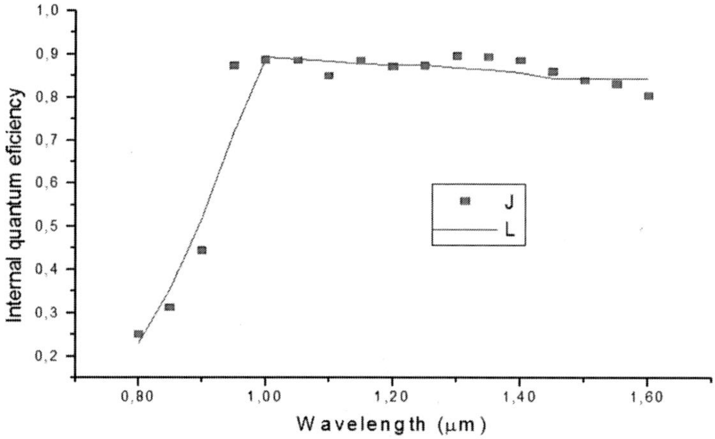

Figure 13. Internal quantum efficiency of photodiode HAM experimental values (dots) and fitted values (solid line) according to the model shown below.

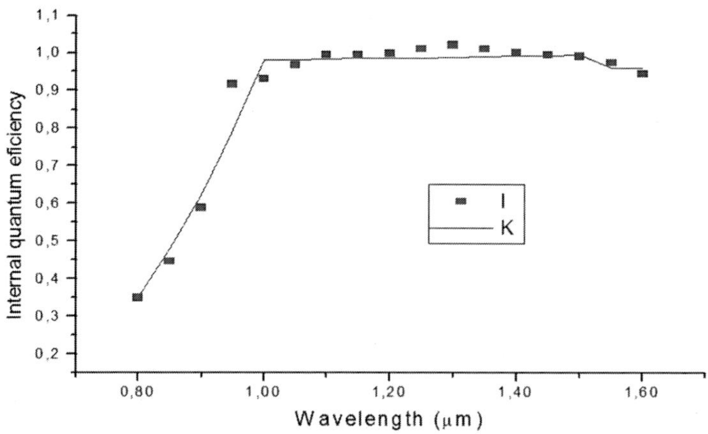

Figure 14. Internal quantum efficiency of photodiodes GPD experimental values (dots) and fitted values (solid line) according to the model shown below.

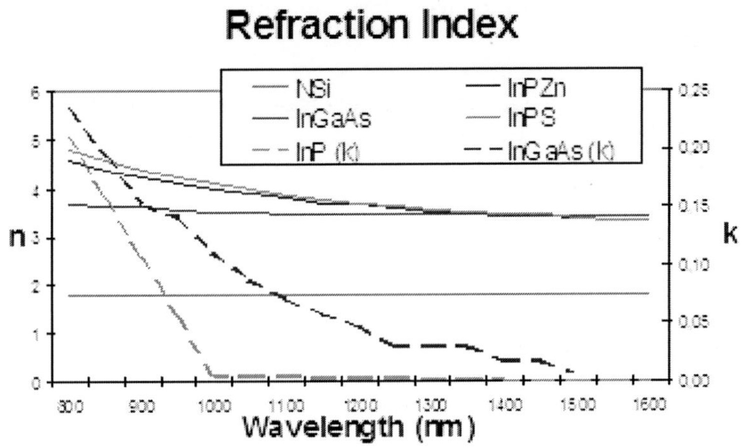

Figure 15. Materials' refraction Index

Thickness of the different layers is obtained by a nonlinear fitting of the experimental values of reflectance to a multilayer model. The model did not worked out well for photodiode POL, then results are not given for it. Perhaps its structure is very different from that of figure 12. Table 3 shows the thickness obtained from the fitting for photodiodes Ham and GPD. The last layer, the deeper one regarding light absorption, was considered to be infinite.

Table 3. Thickness of layers of InGaAs photodiodes

Photodiode	NSi	InP (Zn)	InGaAs
HAM	162.17nm	1213.35nm	1593.2nm
GPD	159.99nm	1200.54nm	1536.7nm

12. CONCLUSIONS

Silicon photodiodes in the visible up to 950 nm and InP/InGaAs photodiodes in the NIR up to 1.6 µm are widely used for optical radiation measurements in many different applications because of their good radiometric properties. They have got high internal quantum efficiency, therefore they are very useful for realizing spectral responsivity scales.

Perhaps in a near future a model be developed for the internal quantum efficiency of InP/InGaAs photodiode as it was done for the silicon, so that its responsivity may be accurately known in their spectral interval of response. Some more work is also needed to know the structure of the device and improve the fitting of reflectance via a multilayer model.

REFERENCES

1. Ferrero, J. Campos, A. Pons, and A. Corrons "New model for the internal quantum efficiency of photodiodes based on photocurrent analysis".*Applied Optics*, Vol. 44, Issue 2, 208-216 (2005).

2. Gentile T.R., Houston J.M., and Cromer C.L. "Realization of a scale of absolute spectral response using the National Institute of Standards and Technology High-accuracy Cryogenic Radiometer. *Appl. Opt.* 35 4392-403 (1996).

3. J. Campos, A. Corróns, A. Pons, P. Corredera, J.L. Fontecha and J.R Jiménez "Spectral responsivity uncertainty of silicon photodiodes due to calibration spectral bandwidth". *Meas. Sci. Technol*, 12, 1936-1921, (2001).

4. J. Campos, A. Corróns, and A. Pons, "Response uniformity of silicon photodiodes". *Applied Optics*, 27, 24, 5154-5156, (1988).

5. E.F. Zalewski and J. Geist, "Silicon photodiode absolute spectral response self-calibration". *Applied Optics*, 19, 8,1214-1216, (1980).

6. R. Schaefer, E.F. Zalewski, and Jon Geist, "Silicon detector nonlinearity and related effects". *Applied Optics*, 22, 8, 1232-1236, (1983).

7. E.F. Zalewski and C.R. Duda"Silicon photodiode device with 100% external quantum efficiency". *Applied Optics*, 22,18, (1980).

8. J. Campos, A. Pons and P. Corredera., "Spectral responsivity scale in the visible range based on single silicon photodiodes. *Metrologia* 40 S181-S184. (2003)

9. Kholer R., Goebel R. and Pello R. "Results of an international comparison of spectral responsivity of silicon photodetectors". *Metrologia* 32 463-8 (1995).

10. E.F. Zalewski and J. Geist "Silicon photodiode absolute spectral response self-calibration". *Appl. Opt.*, Vol. 19, No. 8 pag. 1214-1216. (1980).

11. Born M. and Wolf E. 1989 "Principles of Optics. Electromagnetic Theory of Propagation, Interference and Diffraction of Light". 6th ed. *(Oxford: Pergamon)* p 633

12. Haapalinna A., Karha P. and Ikonen E. "Spectral reflectance of silicon photodiodes". *Appl. Opt.* 37 729-32 (1998).

13. Palik E.D. 1985 Handbook of Optical Constants of Solids. *(New York: Academic Press)*

14. Geist J. and Baltes H. "High accuracy modeling of photodiode quantum efficiency". *Appl. Opt.* 28 3929-39 (1989).

15. Werner L, Fischer J, Johannsen U. and Hartmann J. "Accurate determination of the spectral responsivity of silicon trap detectors between 238 nm and 1015 nm using a laser-based cryogenic radiometer". *Metrologia* 37 279-84, (2000).

16. O. Wada, H. Hasegawa, *InP-Based materials and devices : physics and technology. New York. John Wiley & Sons, 1999.*

17. P. Corredera, M.L. Hernanz, M. González-Herráez, J. Campos "Anomalous non-linear behaviour of InGaAs photodiodes with overfilled illumination". *IOP Metrología* 40, S181-S184, (2003).

18. P. Corredera, M.L. Hernanz, J. Campos, A. Corróns, A. Pons and J.L Fontecha. "Comparison between absolute thermal radiometers at wavelengths of 1300 nm and 1550 nm". *IOP Metrología*. 37. 237-247, (2000).

19. J M Coutin, F. Chandoul, J. Bastie,"Characterization of new trap detectors as transfer standards". *Proceedings of the 9th international conference on new developments and applications in optical radiometry*, (2005).

20. A.L. Muñoz Zurita, J. Campos Acosta, A. PonsAglio. A.S. Shcherbakov. "Medida de la reflectancia de fotodiodos de InGaAs/InP". *Óptica Pura y Aplicada (Spain)*, 40(1), 105-109 (2007).

21. Campos, J. Fontecha, J. Pons, A. Corredera, P. Corróns, A., Measurement of standardaluminiummirrors, reflectance versus light polarization.*Measurement Science and Technology* Volume 9, Issue 2, February 1998, Pages 256-260.

22. A.L. Muñoz Zurita, J. Campos Acosta, A. S. Shcherbakov, A. Pons Aglio, "Differences of silicon photodiodes reflectance among a batch and by ageing". *Optoelectronics Letters*. 4(5), (2008).

23. Ana Luz Muñoz Zurita1, Joaquín Campos Acosta2, Ramón Gómez Jimenez1, Rodrigo Uribe Valladares1. "AN ABSOLUTE RADIOMETER BASED ON In PPHOTODIODES". *Proc. of SPIE*. Vol. 7726, pag. 772628-1- 772628-6.

12

NOISE PERFORMANCE OF TIME-DOMAIN CMOS IMAGE SENSORS

Fernando de S. Campos[1], José Alfredo C. Ulson, José Eduardo C. Castanho and Paulo R. Aguiar

[1] Univ. Estadual Paulista "Júlio de Mesquita Filho" (UNESP) – Bauru campus, Brazil

1. INTRODUCTION

Temporal noise is the main disadvantage of CMOS image sensors when compared to charged couple devices (CCDs) sensor. The typical 3T active pixel sensor (APS) architecture presents as main noise sources the photodiode shot noise, the reset transistor and follower thermal and shot noise, the amplifier thermal and 1/f noise, the column amplifier thermal and reset noise (Zheng, 2011; Brouk, 2010; Jung, 2005; Tian, 2001; Derli, 2000; Yadid-Pecht, 1997). In order to reduce the APS noise several approaches have been proposed in the literature. Some of these approaches are the use of high gain preamplifiers, correlated multiple sampling (CMS) and low bandwidth column-parallel single slope A/D converters (Sakakibara, 2005; Kawai, 2004; Suh, 2010; Lim, 2010; Yoshihara, 2006; Chen, 2012). However, APS in time domain has as advantage to show lower source of noise since it is composed only by a photodiode, a reset transistor and a voltage comparator. It shows as noise source only the reset transistor and the photodiode. Therefore, in principle, APS in time domain may presents lower overall noise.

The only two main noise source of APS in time domain are the reset noise and the integration noise. The source of reset noise is the incomplete reset operation. Tian et al. 2001, show that APS operates usually with incomplete reset operation. The incomplete reset operation originates a random reset voltage that varies from frame to frame as a source noise. It have been found that the reset noise is$kT/2C\ ph$. During the integration period, the photodiode shot noise predominates generating a integration noise that is a function of the integration time, the photocurrent and the dark current. Altought the reset noise is the same to APS in voltage domain and in time domain, the integration noise must present different behavior in both approach. We show that while the integration time increases at higher photocurrents in the voltage domain approach, the integration time is approximately constant at time domain approach. In this chapter the conventional frequency domain noise is not used. Instead, a temporal analysis is presented as proposed by (Tian, 2001).

2. CMOS ACTIVE PIXEL SENSOR (APS) FUNDAMENTALS

2.1. Aps Operation In Voltage Domain

Figure 1 shows a typical architecture of a conventional 3T CMOS active pixel. This pixel comprising a reset transistor M1, a source follower M2 and a line select transistor M3. When the line select is activated the select transistor is on and the source follower transmits the signal to the column bus keeping the photodiode isolated. After, the column signal at active load is amplified and converted to digital.

The conventional APS pixel of Figure 1 presents two operation stages: a reset period and a integration period (Figure 2). During reset period the reset transistor is on and the select transistor is off. In this period the photodiode is reversely charged to V_{dd}. After, the reset transistor is turned off initiating the integration period. In this period the photocurrent discharge the photodiode during an interval time called integration time. At the end of integration time the signal is readout externally by activation of select transistor.

The fixed-pattern noise (FPN) is the non-uniformity introduced in image due to parameters variation from pixel to pixel. It is one of main disadvantage of APS when compared to CCDs. In general, the FPN can be reduced by applying double sampled correlated (CDS). Figure 3 shows a simple circuit that can be used to implement CDS. The CDS operation is comprised by three steps; (1) sample and hold the reset signal, (2) sample and hold the signal after the integration time and (3) subtraction of signals of steps (1) and (2).

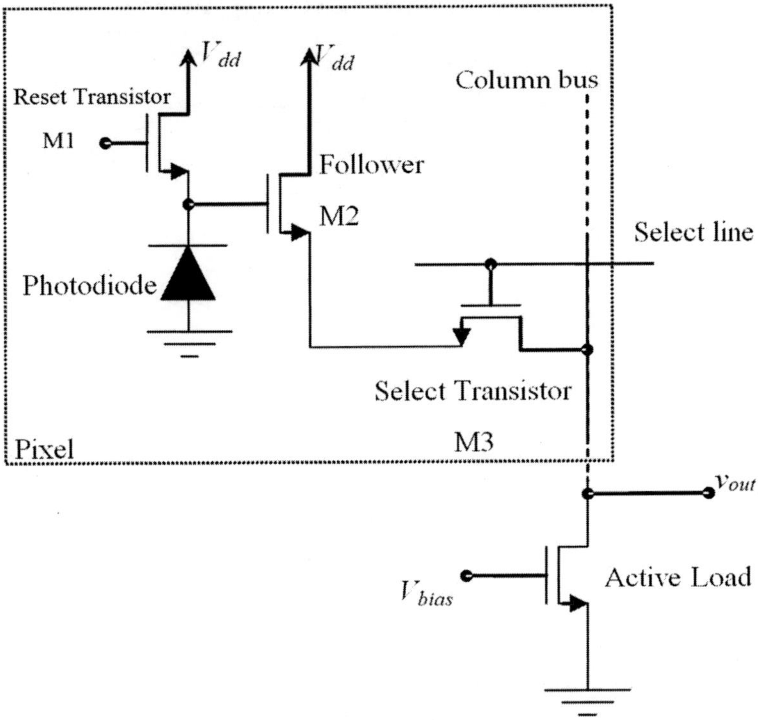

Figure 1. Diagram showing the conventional 3T APS architecture.

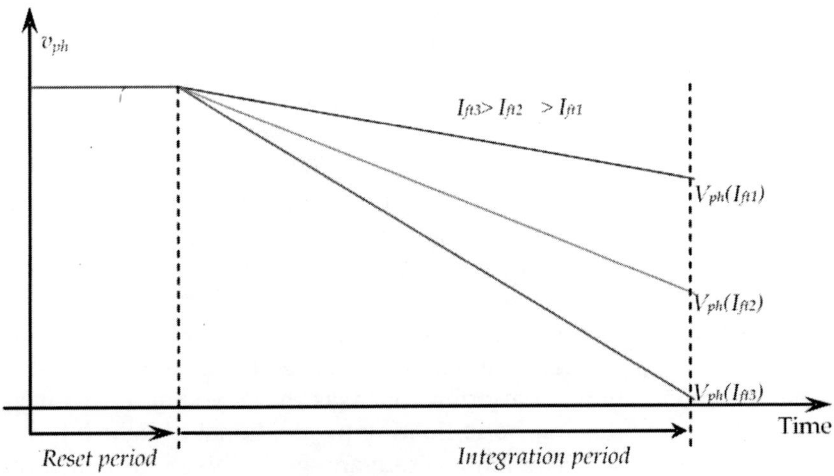

Figure 2. APS operation in voltage domain.

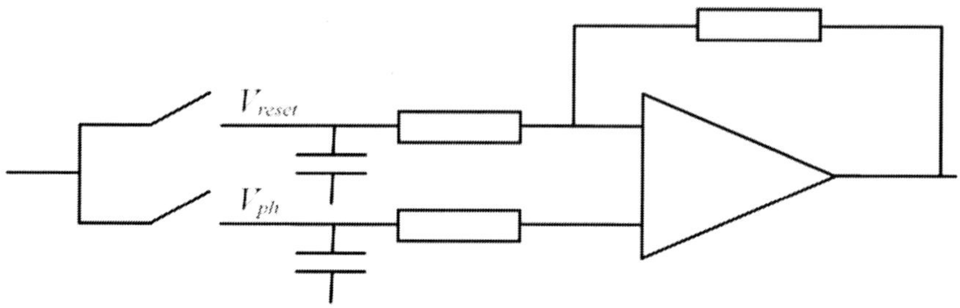

Figure 3. Simple CDS circuit.

The photogate is another type of CMOS photodetector widely used (Fujimori, 2002; Sccherback, 2003; Mendis, 1997). The photogate is composed by a MOS capacitance, a pass transistor and a floating diffusion as shown in Figure 4 (Fossum, 1997). The photogeneration occurs in the depletion region of the MOS capacitor. The photogate operation can be separated into four stages (i) integration, (ii) the floating diffusion reset, (iii) transfer the load to the MOS capacitor floating diffusion and (iv) reading of the floating node voltage signal. During the integration period, the terminal port (PG) of the MOS capacitance is set at V_{dd} and the carriers generated in the depletion region below the gate terminal (PG), are separated by the electric field junction metal-oxide-semiconductor. The polarization of the transfer terminal TX at low level isolates the load MOS capacitor node holding the floating charge under the gate region. The reset operation of the floating node consists on drive the transistor reset loading the floating node voltage V_{dd}. After resetting the floating node, the voltage of the transfer terminal TX is increased and the port terminal PG voltage MOS capacitor is reduced.

The charge transfer occurs with the polarization inversion of the TX and PG. The inversion of the potential resulting from PG and TX polarization cause displacement of charges stored in the region below the gate terminal (highest potential) toward the floating diffusion region (lower potential). After transfer, the transfer terminal voltage TX is reduced again isolating the MOS capacitor of the floating node FD and the loads are stored in the floating node. The new charge balance in FD leads to the floating node voltage variation that was initially charged with V_{dd}. The variation in voltage at node floating proportional to light intensity can be read externally. The main disadvantage of this photodetector is the lowest quantum efficiency, particularly in the blue region of the spectrum due to the reduction of the luminous flux caused by absorption of photons in the upper layer of polysilicon. In general, the circuitry for reading the signal APS systems with phototogate or photodiodes are the same or similar and in both cases are needed to readout the voltage of the photodetector.

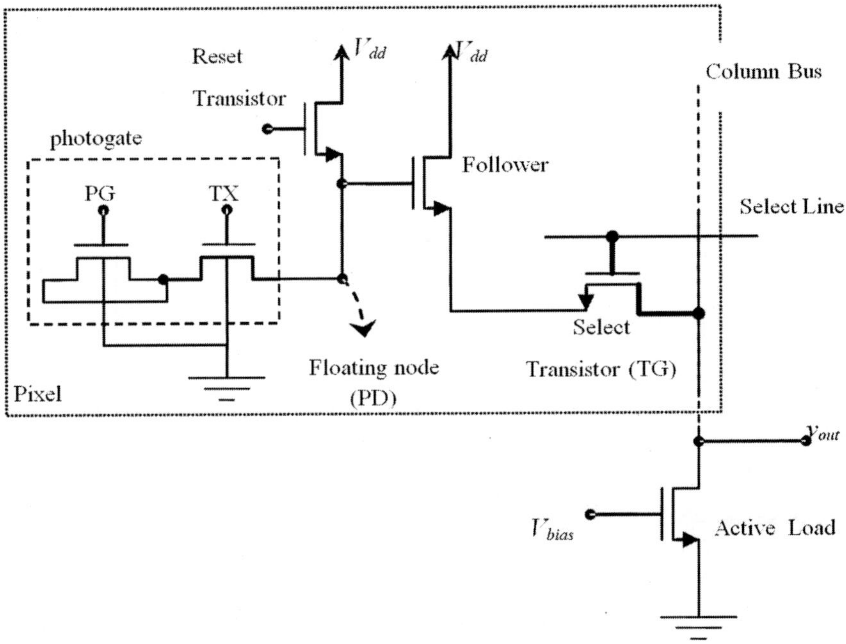

Figure 4. Active Pixel Photogate (PG-APS).

The logarithmic is another important type of CMOS Active Pixel (Figure 6). The logarithmic pixel is particularly attractive in applications that require image capture with high dynamic range (Kavadias, 2000; Joseph, 2002; Choubey, 2006). The logarithmic photoresponse allows you to capture light intensities in ranges of 6 orders of magnitude. However, the logarithmic pixel presents as disadvantages, high fixed pattern noise (FPN), low signal to noise ratio and a small swing of the output voltage. Furthermore, the logarithmic pixel requires long time to reach steady state at low light intensities. The pixel logarithmic is composed by a photodiode connected in series with a MOSFET (Figure 6). The output voltage reaches steady state when the MOSFET current becomes equal to the photocurrent ($I_D = I_{ph}$). The MOSFET operates in series and the gate and drain terminals connected directly ($V_{DS} = V_{GS}$) such that the given output voltage is given by $V_{out} = V_{dd}\text{-}V_{GS}$. Due to low values of photocurrent, the MOSFET operates in the weak inversion region in which the voltage is a logarithmic function of current. Therefore, the output voltage varies logarithmically with the luminous intensity.

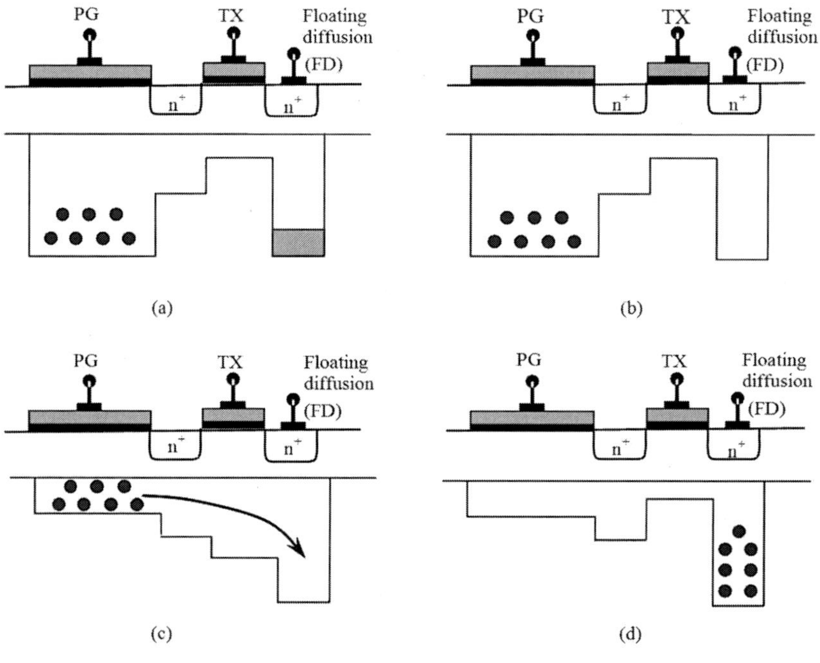

Figure 5. Photogate Operation Stages (potential diagram) (a) integration (b) *reset* (c) transfer (d) readout.

2.2. Aps Operation In Time Domain

Figure 4 shows the architecture of a typical CMOS APS which operates in the time domain. The pixel comprises a photodiode, a reset transistor, a voltage comparator and an 8-bit counter. The operation of the photodiode two basic steps, and integration reset as shown in Figure 5. In the time domain the incident light intensity is related to the time of discharge of the photodiode. The voltage of the photodiode during the integration period is compared with a reference voltage for measuring the time of discharge voltage of the photodiode. At the time the photodiode voltage falls below the reference voltage the counter count for storing the time the voltage of the photodiode lead to vary from the *Vdd*to the reference voltage *Vref*

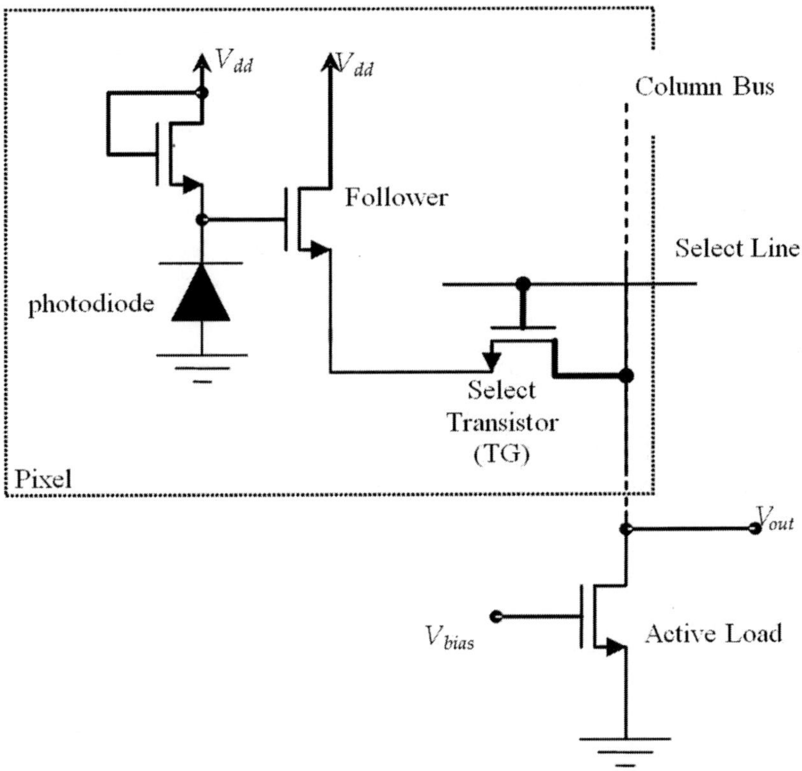

Figure 6. Logarithmic Active Pixel (LOG-APS).

The disadvantage of the APS operating in the time domain is the low fill factor. The fill factor is low due to the integration of the voltage comparator of the 8 bits counter per pixel. Alternatively to the low fill factor APS operating in the time domain in Figure 7, Fields et. al. proposed the method of reading multisampling in the time domain (Campos, 2008). Figure 9 shows a typical architecture of an APS multisampling operating in the time domain. The reading method consists in sampling the comparison result at time intervals. This method makes it possible to integrate the comparatorand the counter by column and therefore outside the pixel, reducing the number of integrated transistors per pixel and increasing the fill factor significantly.

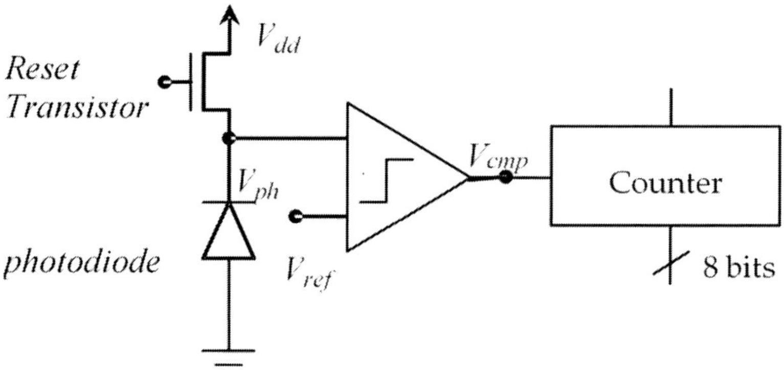

Figure 7. Typical APS architecture operating in time domain.

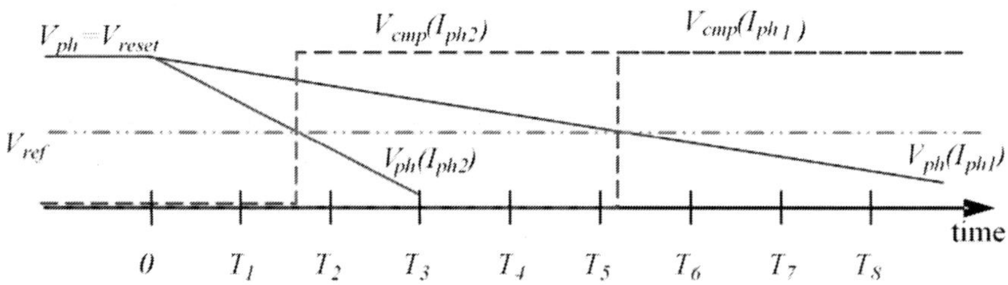

Figure 8. Main waveforms of APS operating in time domain.

3. TEMPORAL NOISE

In this section we present the temporal noise analysis proposed by (Tian, 2001). Only the noise sources presented in time-domain APS are showed: the reset noise, and integration noise. During the reset period the charge time of photodiode is usually greater than the period. Also, the charge time in the reset period is a function of light intensity incident. Therefore, the voltage at end of reset period and beginning of integration time varies generating a random variation of voltage measured. This variation is known as reset noise. According Tian et al, the quadratic mean voltage of reset noise is given by

$$\overline{V_n^2} \cong \frac{kT}{2C_{ph}} \tag{1}$$

where k is the Boltzmann constant; T is the temperature in Kelvin and C_{ph} is the photodiode capacitance. Figure 10 shows the RMS reset noise voltage as a function of the photodiode capacitance at T=300K. For capacitances from 20fF to 100fF the RMS reset noise voltage is about a few hundred of milivolts.

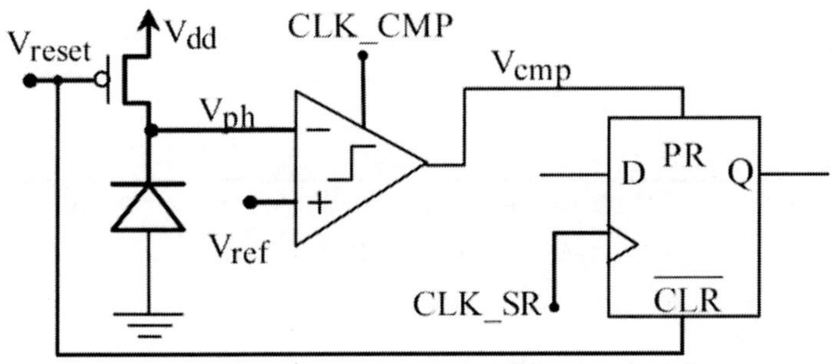

Figure 9. Typical APS multi-sampled in time-domain architecture.

During the integration period the integration noise is composed by the shot noise related to the photocurrent and dark current. The quadratic mean voltage of the integration noise is given by

$$\overline{V_n^2(t_{int})} = \frac{q(i_{ph} + i_{dark})}{C_{ph}^2(v_{ph}(0))} t_{int} \left(1 - \frac{1}{2(v_{ph}(0) + \phi)} \frac{(i_{ph} + i_{dark})}{C_{ph}(v_{ph}(0))} t_{int}\right)^2 \tag{2}$$

where q is the elementary charge, *iph* is the photocurrent, *idark* is the dark current, *vph* is the photodiode voltage at integration time beginning, *Cph* is the photodiode capacitance and *tint* is the integration time (Tian, 2001). Figure 11 shows the RMS integration noise voltage considering *Cph*=30pF, *idark*=2fF,*vph*(0)=3V and *tint*=30ms. The RMS voltage integration noise is about a few milivolts for photocurrents in the range of 0.1pA to 1pA.

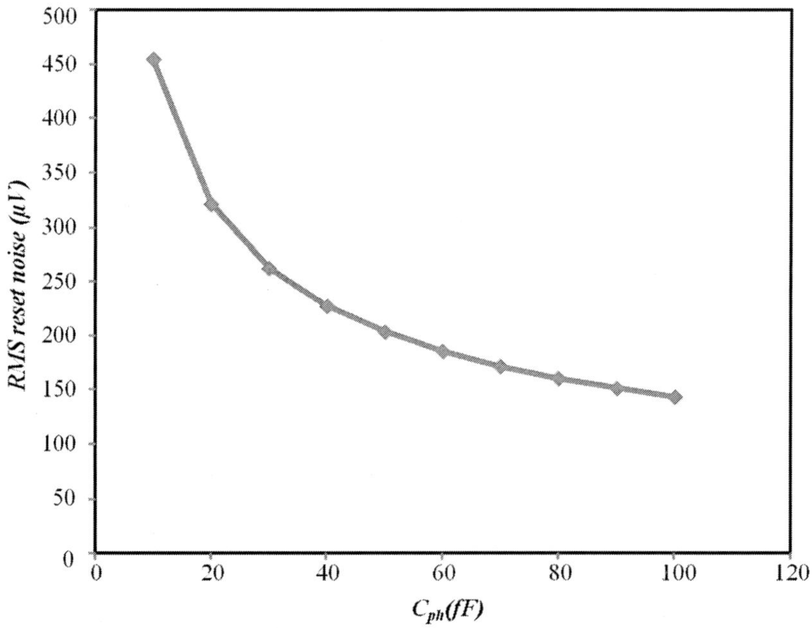

Figure 10. RMS reset noise voltage.

For APS operating in voltage domain, the follower transistor and select transistor of conventional APS also contributes to the total noise in APS, however, they are not presented in APS operating in time-domain and will be ignored in this analysis.

4. TEMPORAL NOISE IN TIME DOMAIN

As the APS in time domain pixel architecture is composed by a transistor reset, a voltage comparator and the photodiode, the main sources of noise are the reset noise and the integration noise as described in the last subsection. The reset operation of APS time domain is the same that the reset operation of APS voltage domain and, thus, the reset noise in time is the same given by equation (3). However, the integration time is different in time domain since the integration time is different for different values of photocurrent (see eq. (2) and (4)). Substituting eq. (2) in eq. (4), the integration noise in time domain is given by

$$\overline{V_n^2} = \frac{q(i_{ph} + i_{dark})}{C_{ph}(v_{ph}(0))} \left(\frac{(v_{ph}(0) - v_{ref})}{i_{ph}} \right) \left(1 - \frac{1(i_{ph} + i_{dark})}{2(v_{ph}(0) + \phi)} \left(\frac{(v_{ph}(0) - v_{ref})}{i_{ph}} \right) \right)^2 \quad (3)$$

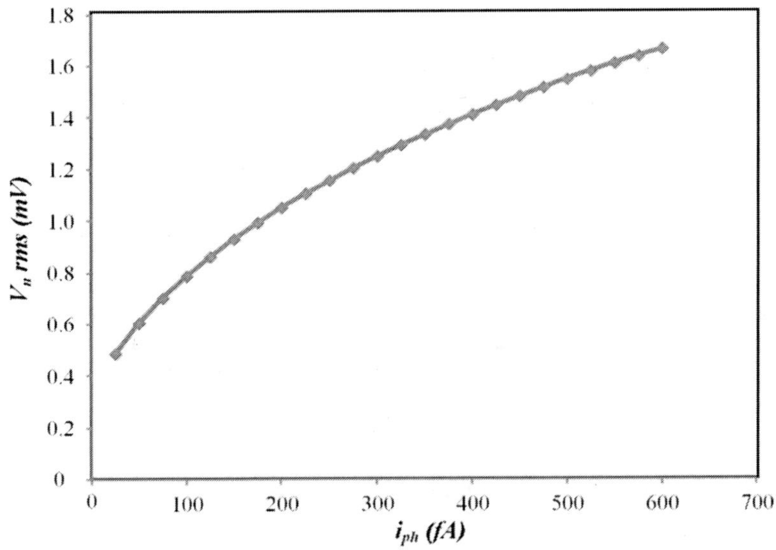

Figure 11. RMS integration voltage noise.

Figure 12 shows the integration noise voltage given by eq. 5 assuming *Vref*=1.5V the same values ofFig. 11. As one can see the integration noise is approximately constant for *iph>>idark* and it is a *Vref*function where the RMS integration voltage noise decreases as the reference voltage increases.

The voltage noise must reflect in a time noise on the comparison time given by equation (2). Assuming the equivalent circuit of Figure 10, the comparison time given by eq. (2) can be written as

$$t_d \cong \frac{\left(V_{reset} - V_{ref} + \overline{V_n}\right)}{i_{ph}} C_{ph}\left(v_{ph}(0)\right) \quad (4)$$

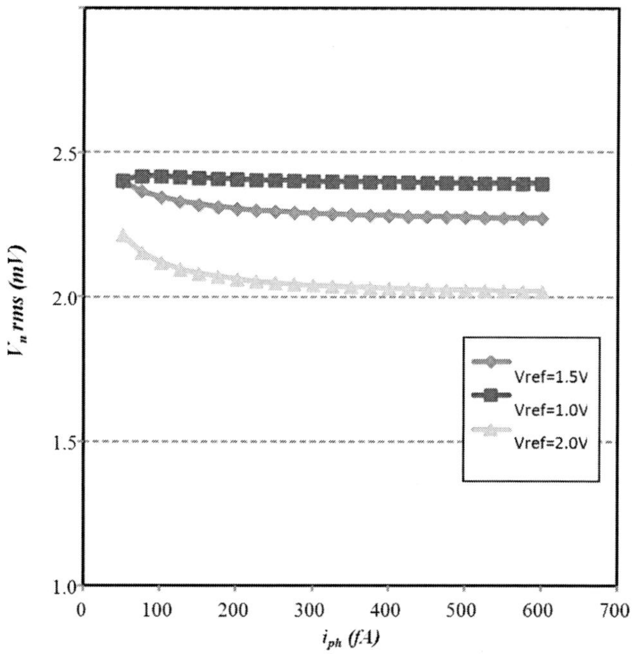

Figure 12. RMS integration voltage noise for APS operating in time-domain.

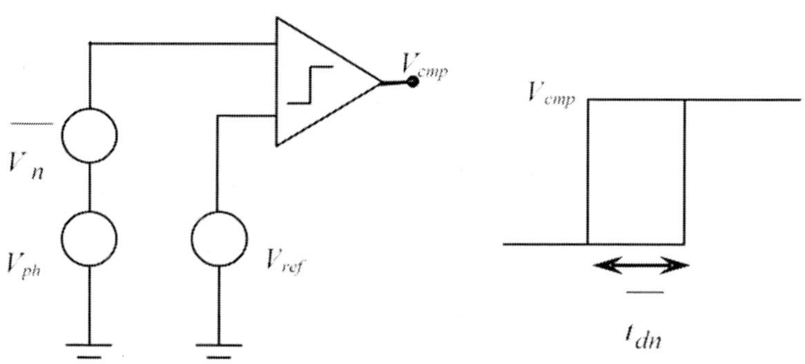

Figure 13. Noise in time-domain.

Manipulating equation (6) the discharge time with noise can be written as

$$t_d \cong T_d + \overline{t_{dn}} = \frac{(V_{reset} - V_{ref})}{i_{ph}} C_{ph}\left(v_{ph}(0)\right) + \frac{\overline{V_n}}{i_{ph}} C_{ph}\left(v_{ph}(0)\right) \tag{5}$$

where *Td* is the discharge time without noise and *tdn* is the noise in time. However the signal-to-noise (SNR) ratio in time domain can be defined as

$$SNR_{time} = \frac{T_d}{\overline{t_{dn}}} = \frac{(V_{reset} - V_{ref})}{\overline{V_n}} \tag{6}$$

Figure 14 shows the *SNRtime* assuming *Vreset*=3Vfor the three cases of Figure 12. The SNR drops at lower photocurrents values while it keeps constant at higher photocurrent values. Also, it is possible to note that the SNR values increase slightly as the reference voltage increases. However, one can note that for low reference value as at *Vref*= 1.0V, the SNR is approximately constant for over the photocurrent range.

5. EXPERIMENTAL RESULTS

In this section the results of experiments show the behavior of noise in a multi-sampled time-domain APS proposed by (Campos, 2008). A pixel as shown in Figure 9 was implemented in 0.35μm AMS technology. The measurements were performed using an illuminator (Spectra Physics, with 100W Xenon lamp), optical filters and an integration sphere in a dark room. Figure 15 shows the measurement result for the characteristic discharge time versus illumination intensity, using a constant voltage of *V ref* =1.5V. The analysis of slope in Figure 15 showed that the pixel sensitivity is about 3.4V-cm^2/s-W.

Figure 16 shows the SNR as a function of light intensity obtained from the standard deviation and the medium comparison time obtained from the measurements at *V ref*=1.5V. As expected from theoretical results, the SNR is approximately constant during 3 decades. It presents a SNR average of 54dB while the theoretical results is about 56dB. The SNR drops observed at high light intensities (>10^{+4}W/cm^2) may be related to the slew-rate of comparator that limits its performance at higher frequency operation. However, a better measurement procedure must be developed to caracterize the reference voltage effects and the comparator noise contribuition.

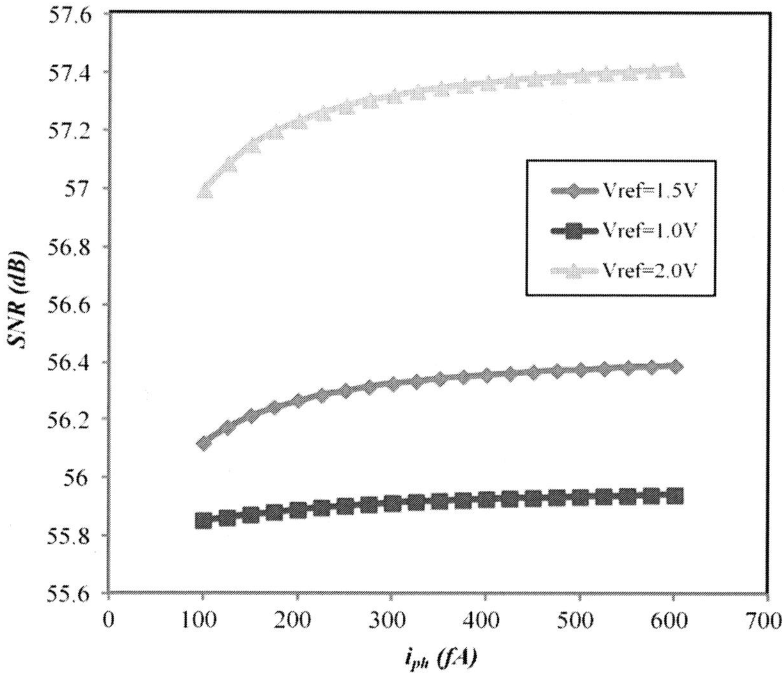

Figure 14. SNR for APS operating in time-domain.

6. CONCLUSIONS

The temporal noise analysis of APS in time domain was presented. Theoretical noise analysis indicated that the noise is constant when APS operates in time domain. The SNR is approximately constant when APS operates in time domain while the literature indicates that the SNR of APS operating in voltage domain drops at lower light intensities. Therefore, the results indicate that the operation in time domain is more suitable for low noise applications. Experimental results are in agreement with theoretical analysis.

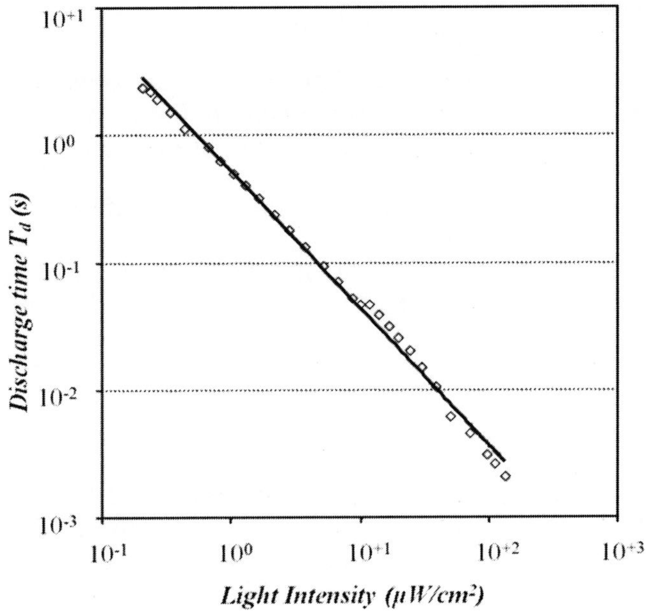

Figure 15. Discharge time.

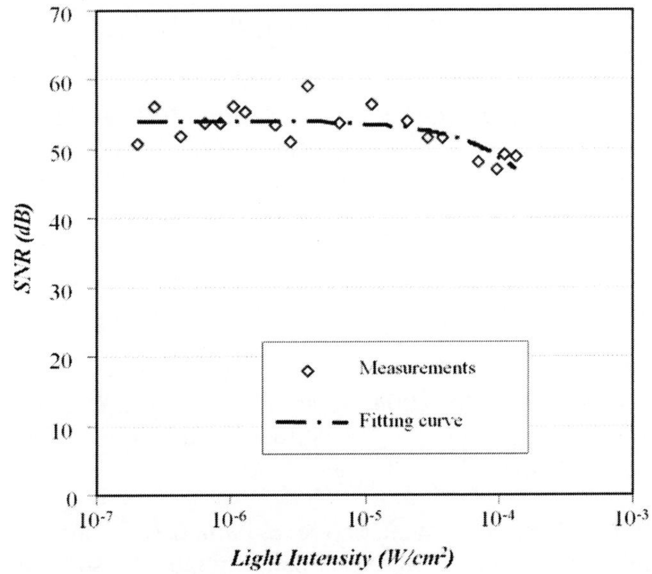

Figure 16. SNR in a time domain APS

REFERENCES

1. I. Brouk, A. Nemirovisky, K. Alameh, Y. Nemirovsky, 2010 Analysis of noise in CMOS image sensor based on a unified time-dependent approach Journal Solid-State Electronics Elsevier 54 1 28 36

2. F. S. Campos, O. Marinov, N. Faramarzpour, F. Saffih, M. J. Deen, J. W. Swart, 2008 A multisampling time-domain CMOS imager with synchronous readout circuit Analog Integrated Circuits and Signal Processing Journal 57 151 159

3. B. Choubey, S. Aoyama, S. Otim, D. Joseph, S. Collins, 2006 An Electronic-Calibration Scheme for Logarithmic CMOS Pixels IEEE Sensors Journal 6 4 950 956

4. Y. Chen, Y. Xu, A. J. Mierop, A. J. P. Theuwissen, 2012 Column-Parallel digital correlated multiple sampling for low-noise CMOS image sensors IEEE Sensors Journal 12 4 793 799

5. Y. Derli, F. Lavernhe, P. Magnan, J. A. Farre, 2000 Analysis and reduction of signal readout circuitry temporal noise in CMOS image sensors for low-light levels IEEE Transactions on Electron Devices 47 5 949 962

6. E. R. Fossum, 1997 CMOS Image Sensors: Electronic Camera-On-a-Chip IEEE Transactions on Electron Devices 44 10 1689 1698

7. I. L. Fujimori, W. Ching-Chun, C. G. Sodini, 2002 A 256x256 CMOS Differential Passive Pixel Imager with FPN Reduction Techniques IEEE Journal of Solid-State Circuits 35 12 2031 2037

8. S. Kavadias, B. Diericks, D. Scheffer, A. Alaerts, D. Uwaerts, J. Bogaerts, 2000 A Logarithmic Response CMOS Image Sensor with On-Chip Calibration IEEE Journal of Solid-State Circuits 35 8 1146 1152

9. N. Kawai, S. Kawahito, 2004 Noise analysis of high-gain low-noise column readout circuits for CMOS image sensors. IEEE Transactions on Electron Devices 51 2 185 194

10. S. K. Mendis, S. E. Kemeny, R. C. Gee, B. Pain, C. O. Staller, Q. Kim, E. R. Fossum, 1997 CMOS Active Pixel Image Sensors for Highly Integrated Imaging Systems IEEE Journal of Solid-State Circuits 32 2 187 197

11. D. Joseph, S. Collins, 2002 Modeling, Calibration, and Correction of Nonlinear Illumination-Dependent Fixed Pattern Noise in Logarithmic CMOS Image Sensor IEEE Transactions on Instrumentation and Measurements 51 5 996 1001

12. C. Jung, M. H. Izadi, M. L. La Haye, G. H. Chapman, K. S. Karim, 2005 Noise analysis of fault tolerant active pixel sensors In proceedings of the 20th IEEE International Symposium on Defect and Fault Tolerance in VLSI Systems 140 148

13. Y. Lim, K. Koh, K. Kim, H. Yang, J. Kim, Y. Jeong, S. Lee, H. Lee, S-H. Lim, Y. Han, J. Kim, J. Yung, S. Ham, Y-T. Lee, 2010 A 1.1e- temporal noise 1/3.2 inch 8 Mpixel CMOS image sensor using pseudo-multiple sampling In proceedings of International Solid-State Circuits Conference (San Francisco, CA, February) 396 397

14. M. Sakakibara, S. Kawahito, D. Handoko, N. Nakamura, H. Satoh, M. Higashi, K. Mabuchi, H. Sumi, 2005 A high-sensitivity CMOS image sensor with gain-adaptive column amplifiers IEEE Journal of Solid State Circuits 50 5 1147 1156

15. I. Sccherback, O. Yadid-Pecht, 2003 Photoresponse Analysis and Pixel Shape Optimization for CMOS active pixel Sensors IEEE Transactions on Electron Devices 50 1 12 18

16. S. Suh, S. Itoh, S. Aoyama, S. Kawahito, 2010 Column-parallel correlated multiple sampling circuits for CMOS image sensors and their noise reduction effects Sensors 10 9139 9154

17. H. Tian, B. Fowler, A. El Gammal, 2001 Analysis of temporal noise in CMOS photodiode active pixel sensor IEEE Journal of Solid-State Circuits 36 1 92 101

18. O. Yadid-Pecht, B. Mansoorian, E. Fossum, B. Pain, 1997 Optimization of noise and responsivity in CMOS active pixel sensors for detection of ultra low light levels In Proceedings of SPIE (San Jose, CA, February 25) 3019 125 136

19. S. Yoshihara, et al. 2006 A 1/1.8-inch 6.4 Mpixel 60 frames/s CMOS image sensor with seamless mode change IEEE Journal of Solid-State Circuits 41 12 2998 3006

20. R. Zheng, T. Wei, D. Gao, Y. Zheng, F. Li, H. Zeng, 2011 Temporal noise analysis and optimizing techniques for 4T pinned photodiode active pixel In proceedings of IEEE International Conference on Signal Processing, Communications and Computing (ICSPCC) 1 5

INDEX

P

Q

R

S

T

U

W

Z